Name ______________________

Directions

Cut out the ladybugs. Work with a partner. Take turns.
Show 4 ladybugs in the top row.
Show the same number in the bottom row.
Then show more ladybugs in the bottom row.
Then show fewer ladybugs in the bottom row.

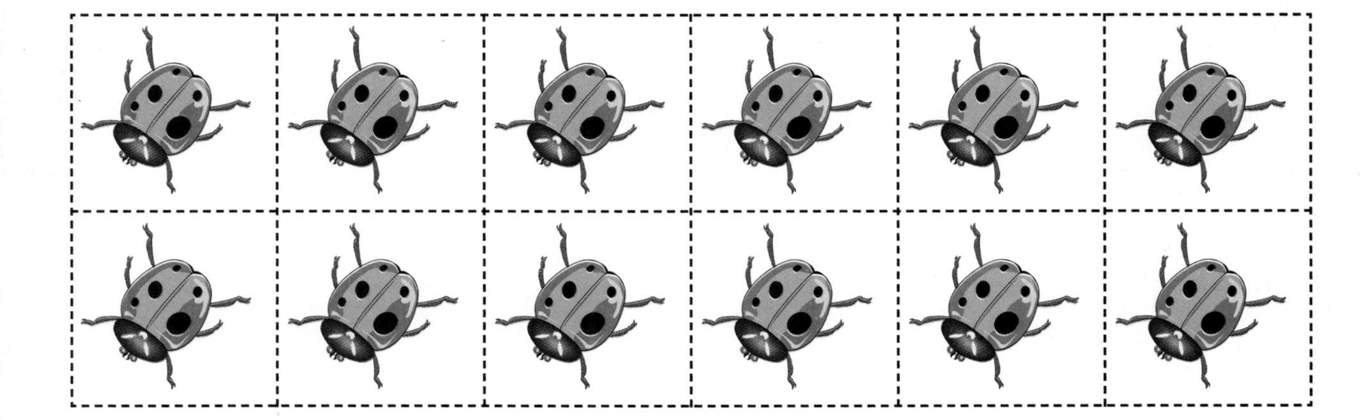

UNIT 1 Vocabulary Words

same	in all +
more	are left −
fewer	equals =

1, 2, 3, 4, 5, 6, 7, 8, 9, 10!
12
FOLD
I learn things, too.
5
CUT
CUT
I learn so much. I am growing up!
16
FOLD
Growing Up
This book belongs to
1

Do you?
6
CUT
FOLD
Wiggle them. Wave them.
Hold them up high.
11
CUT
What is wrong?
2
FOLD
That is the same
of each, you know.
15

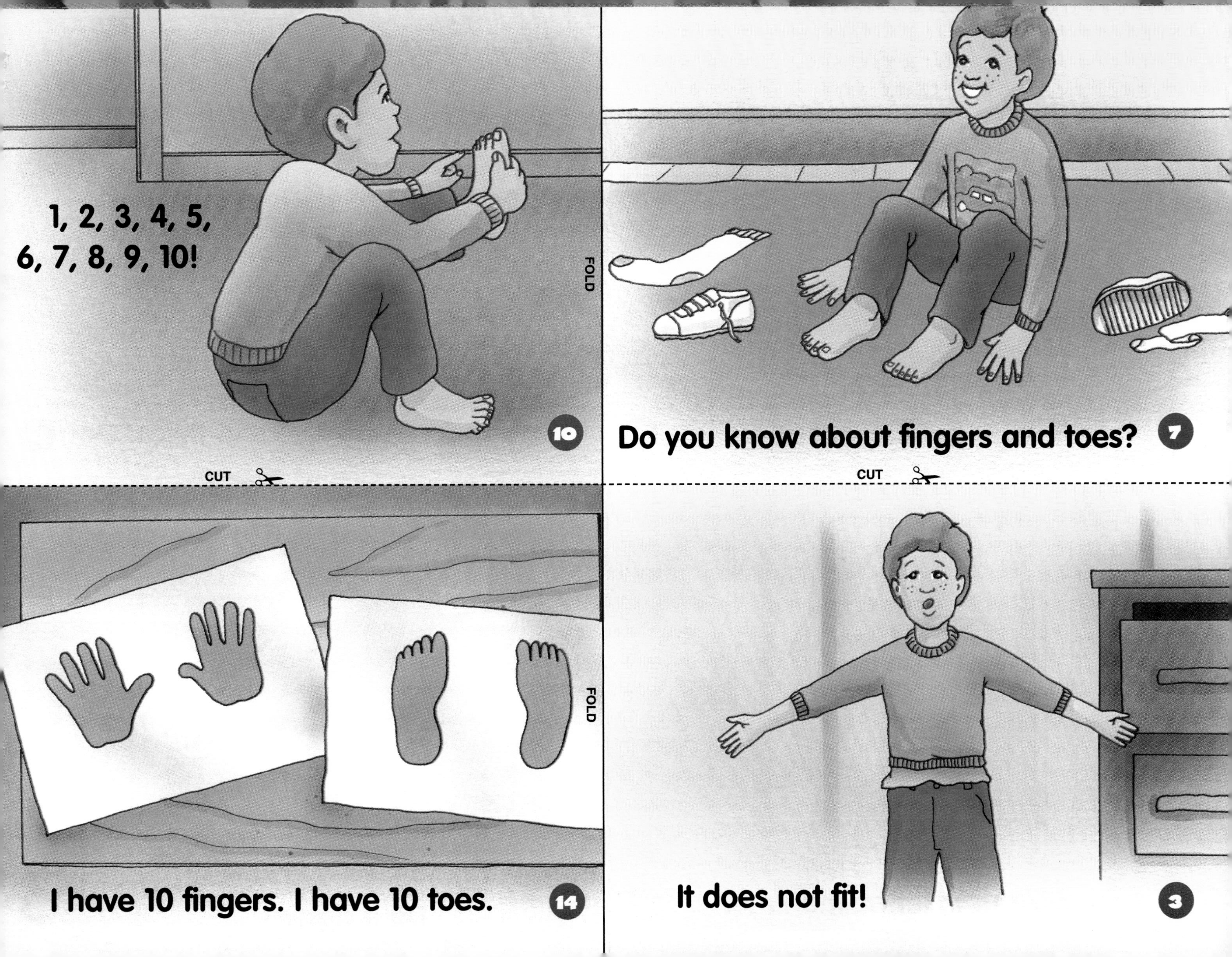
1, 2, 3, 4, 5,
6, 7, 8, 9, 10!
10
FOLD
Do you know about fingers and toes?
7
CUT
CUT
I have 10 fingers. I have 10 toes.
14
FOLD
It does not fit!
3

How many of each do you have?
Do you know?
8
Get ready. Get set. Count them. Go!
9
CUT
CUT
FOLD
FOLD
I am getting taller all the time.
4
Wiggle them. Wave them.
Do it again!
13

Name ______________________________

Directions

Draw lines to match the objects.
The first one is done for you.

1.

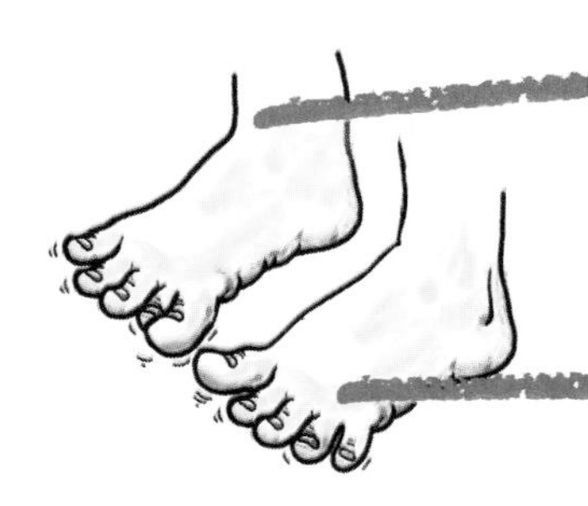

2.

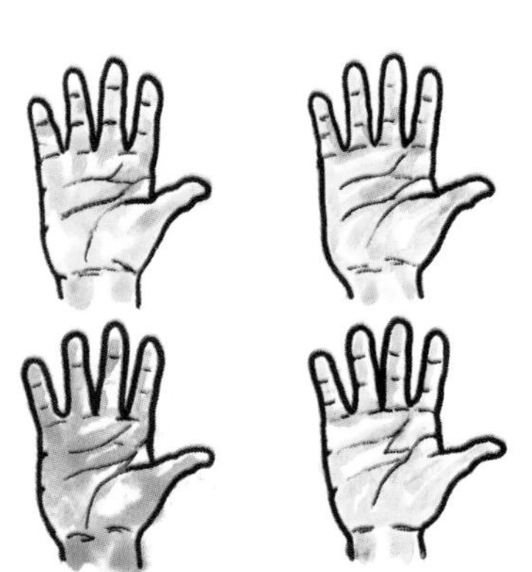

3.

Name ______________________________

Directions

Draw to show the same number of scoops of ice cream.
Then color them.

1.

Directions

Count the objects in the first group. Draw a ring around the same number of objects in the second group.

2.

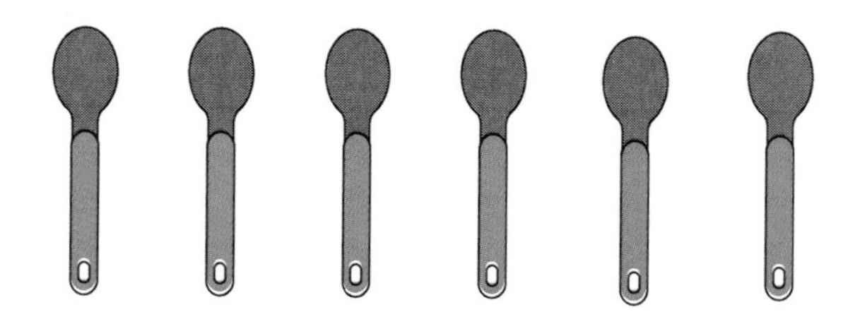

3.

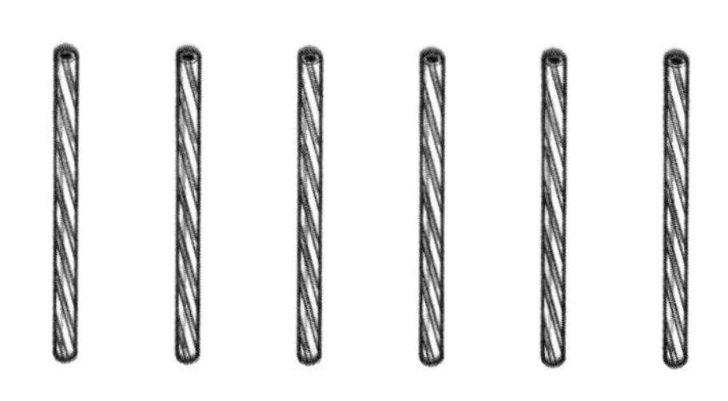

Invite your child to set the table for a family meal. Provide forks and plates. Ask your child to decide whether there are enough forks for plates. Then have your child place a fork next to each plate to show that there is the same number of forks and plates. Try this activity again using glasses.

Name ________________________________

Directions

Draw a ring around the group that shows more objects. The first one is done for you.

1.

2.

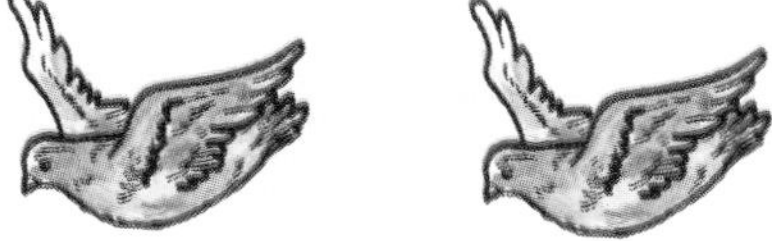
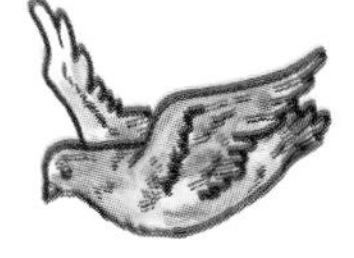

3.

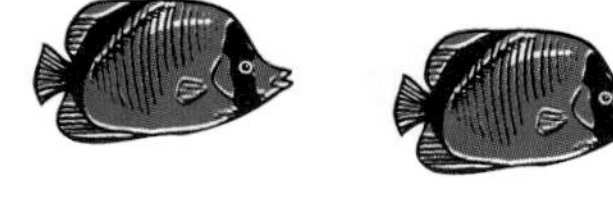

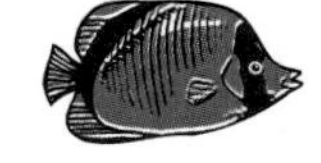

Name

Directions

Draw a ring around the group that shows fewer objects.
The first one is done for you.

1.

2.

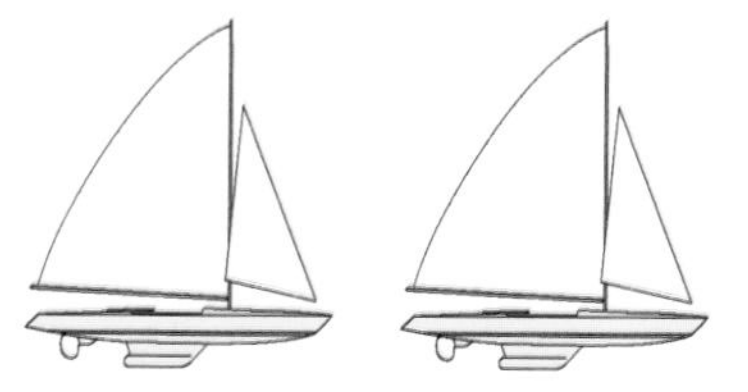

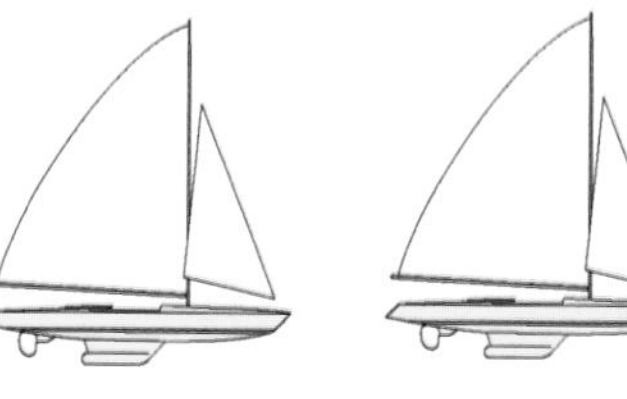

3.

Gather several pairs of shoes. Show your child a group of socks that number more than the shoes. Ask your child to decide whether he or she has the same number, more, or fewer, socks than shoes. Have your child put a sock in each shoe to check his or her answer.

Name

Directions

Cut out each card. Take turns with a partner.
Show a card. Name the coin or coins and the amount.
Have your partner show another card with the same amount.

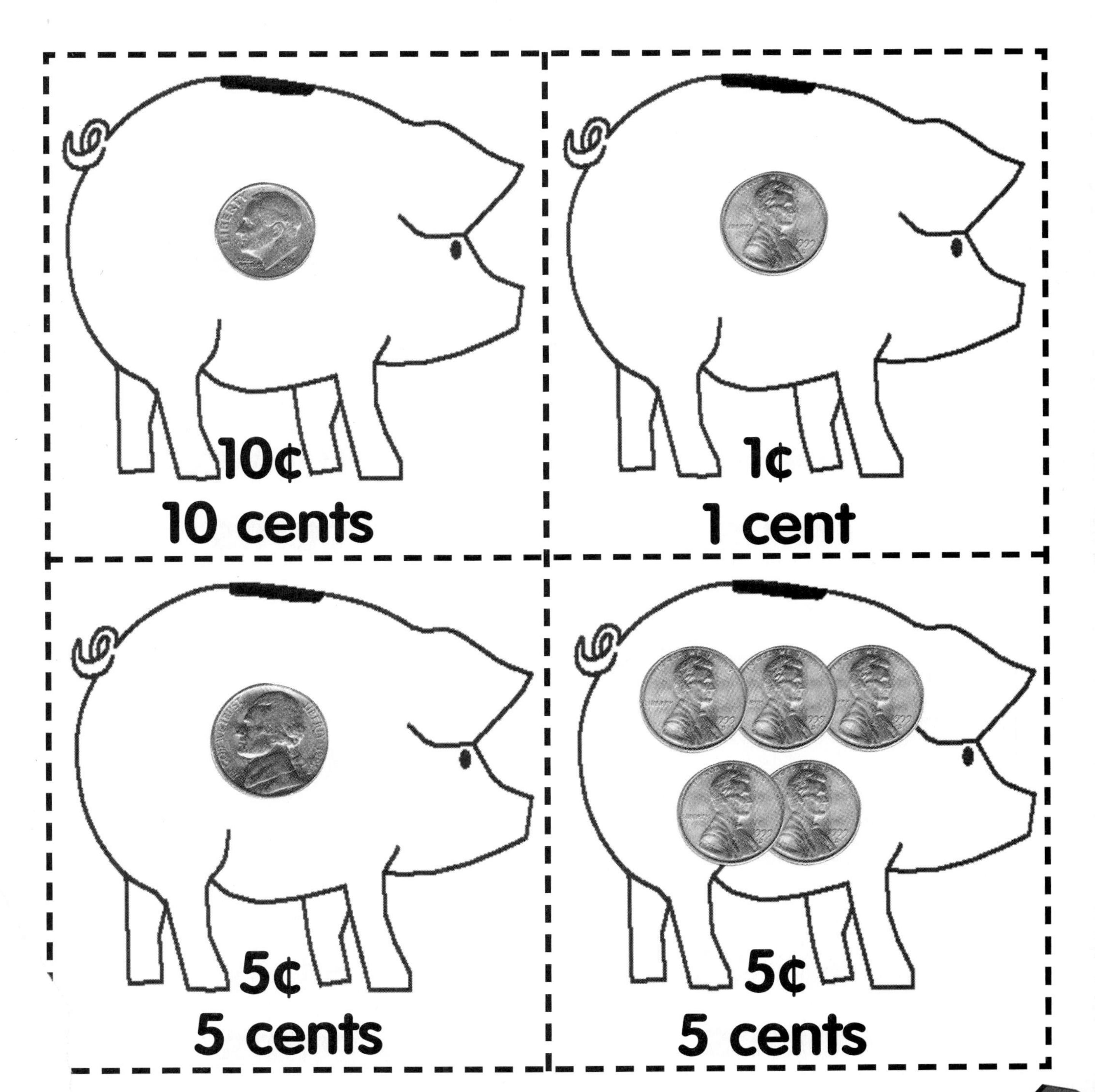

UNIT 2 Vocabulary Words

penny 
nickel
dime 

cents ¢ 
amount 5¢

clock 
time
hour

10¢
10 cents

10¢
10 cents

15¢
15 cents

15¢
15 cents

She says, "Thanks. Have
a nice day."
12
CUT
FOLD
"Get an apple.
It will help you grow."
5
CUT
Señora Verde's store
closes at five.
CLOSED
16
FOLD
Señora Verde's Store
This book belongs to
1

We choose some fruit.
6
CUT
I give her
the money.
FOLD
11
CUT
We know how much money to bring.
2
We are done
just in time.
FOLD
15

Señora Verde tells us
how much to pay.
10
CUT
FOLD
We choose veggies, too.
7
CUT
Then we go on our way.
14
FOLD
Pennies, nickels, and dimes
to buy things.
3

Everything looks
great!
8
Fruits and veggies are good for you.
9
FOLD
CUT
CUT
When we walk up,
Señora Verde says "hello."
4
FOLD
We put them
in a bag.
13

Name ________________________________

Directions

Draw a box around each penny. Draw a ring around each nickel. Mark each dime with an X. Write how many of each coin.

1. How many  s? ______

2. How many  s? ______

3. How many  s? ______

4. How many coins in all? ______

Name ______________________

Directions

Count each group of money. Write the amount.
Draw a ring around the coin that shows the same amount.
The first one is done for you.

1.  5¢  

2.  ______  

3.   ______  

Directions

Draw 2 ways to make 10¢. Draw a ring around the group with fewer coins.

4.

Place pennies, nickels, and dimes in a mixed pile on a table. Have your child sort the coins by type telling you what each coin is worth. Take a nickel and ask your child to show you how many pennies are the same value. Repeat the activity with a dime.

Name ______________________________

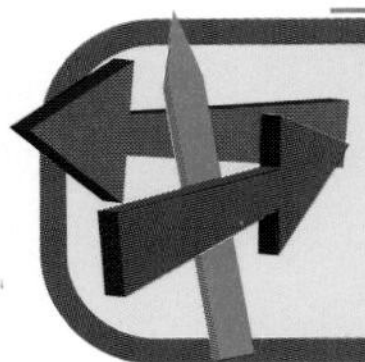

Directions

Look at each pair of pictures. Draw a ring around the activity that takes more time.

1.

2.

3.

Name ______________________________

Directions

Look at each pair of pictures.
Draw a ring around the picture that happens first.

1.

2.

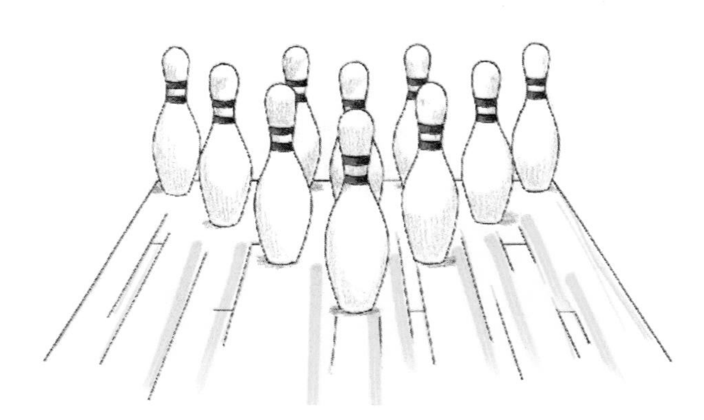

3.

Talk about everyday activities with your child such as getting dressed, walking or riding to school, or watching a favorite television program. Ask your child tell which activity takes more or less time than another. Then have your child think of a typical school day and recite a sequence of 3 or 4 activities in order.

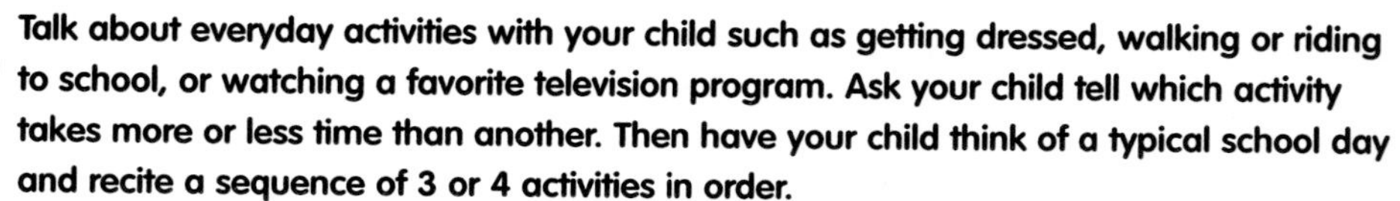

Name ______________________

Directions
Write the missing numbers on the clock.
Then draw hands to show 5:00.

1.

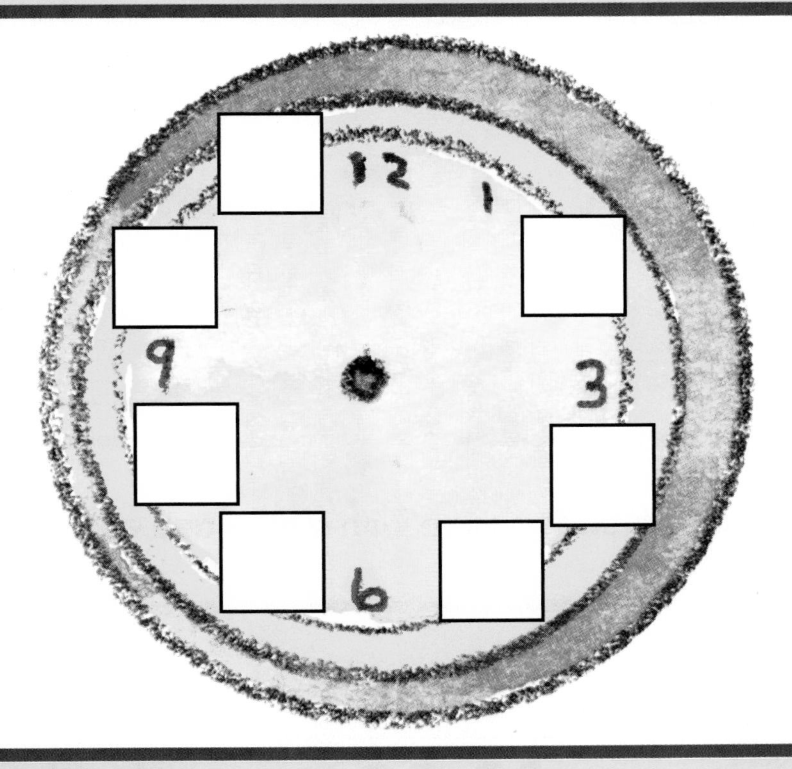

Directions
Look at the clock. Write the time.

2.

___ :00

3.

___ :00

4.

___ :00

Name ______________________________

Directions

Look at the time on each clock. Then draw the hour hand on the clock to show the time.

1. 3:00

2. 10:00

Directions

Draw lines to match each picture with a time and each clock with a time.

3.

10:00

4.

12:00

Ask your child to look at an analog or digital clock and tell you the time several times each day.

Name ______________________________

Directions

Count each group of coins. Write the amount.

1.

_____¢

2.

_____¢

Directions

Look at each toy. Draw a line from the toy to a group of coins needed to buy the toy.

Name ______________________________

Directions

Draw a ring around the picture to answer the question.

1. Which takes more time?

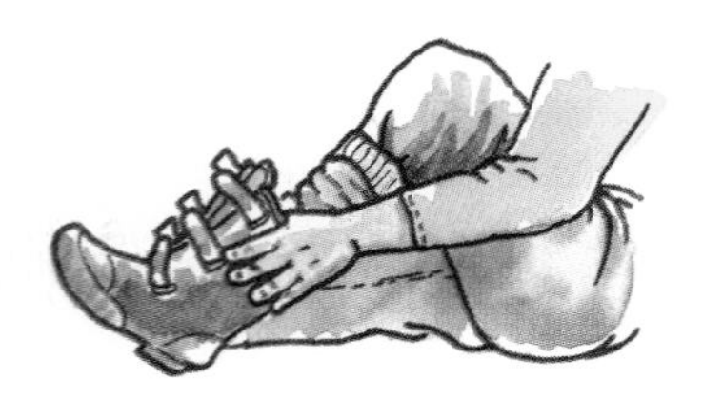

2. Which takes less time?

Directions

Draw a ring around the clock to show when the activity happens

3.

4.

5.

Name

Directions

Cut out the rulers. Work with a partner.
How many cubes high is the fuzzy friend?
How many bears high is it? Talk about it.

cube ruler:

bear ruler:

UNIT 3 Vocabulary Words

long	short	heavier	measure
longest	shortest	lighter	ruler
tall	same	more	
tallest	size	less	

Let's find out.
12
FOLD
CUT
Start with the shortest one.
5
CUT
What do you think of our fuzzy friend parade?
16
FOLD
Fuzzy Friend Parade
This book belongs to
1

The tallest one will go at the other end.
6
How long is the parade?
11
CUT
CUT
FOLD
FOLD
MARCH
Everyone brings a fuzzy friend.
2
We march with our toys.
Hip, hip hooray!
15

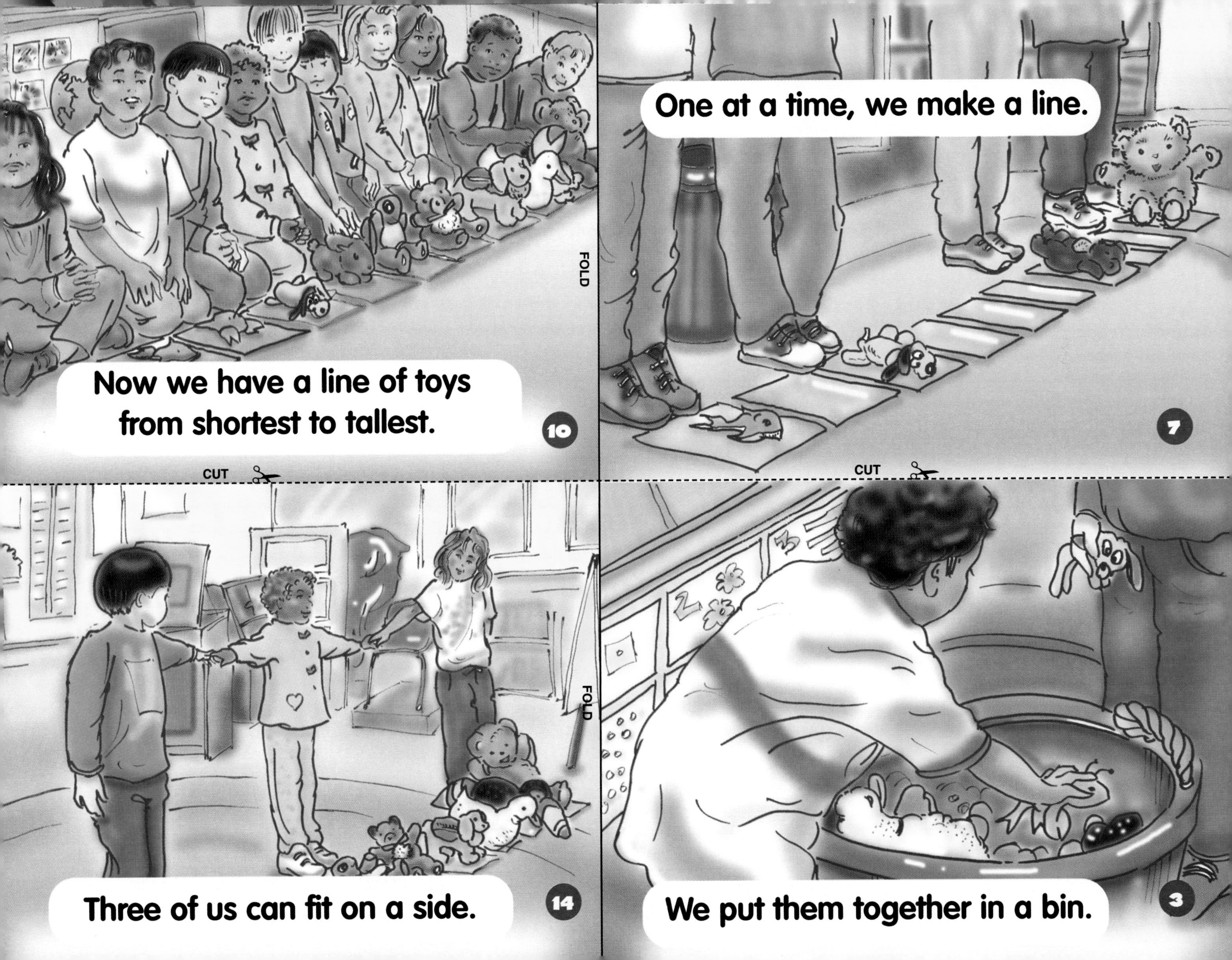
Now we have a line of toys
from shortest to tallest.
10
FOLD
One at a time, we make a line.
7
CUT
CUT
Three of us can fit on a side.
14
FOLD
We put them together in a bin.
3

Oops! The dog is taller than the bear.
8
FOLD
Move the bear down a bit.
9
CUT
CUT
We will line them up.
4
FOLD
We stand with our arms stretched out.
13

Name

Directions
Draw a ring around the objects that are about the same size.

1.

2.

Directions
Work with a partner.
Find 2 or 3 things in the room that are about the same size.
Draw a picture of them.

3.

Name

Directions

Color the tall animal BLUE.
Color the short animal RED.

1.

2.

3.

4.

Help your child understand size. Periodically, hand him or her an object and ask for another object to be found that is the same size. Then ask him or her to find an object that is taller or shorter than your object.

Name ______________________________

LESSON 2 MEASUREMENT

Directions

Draw an X on the longest snake.
Draw a ring around the shortest snake.

1.

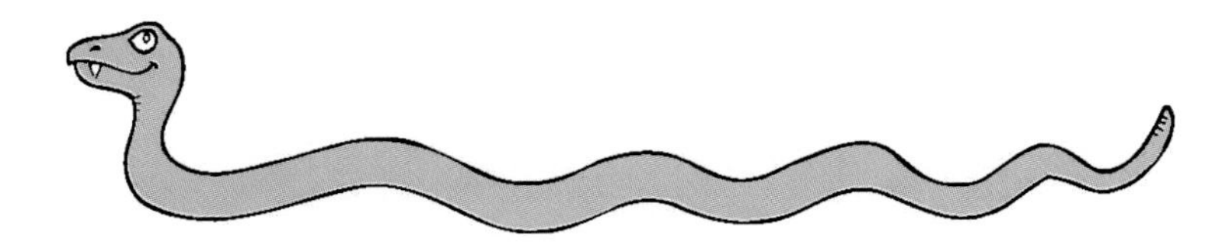

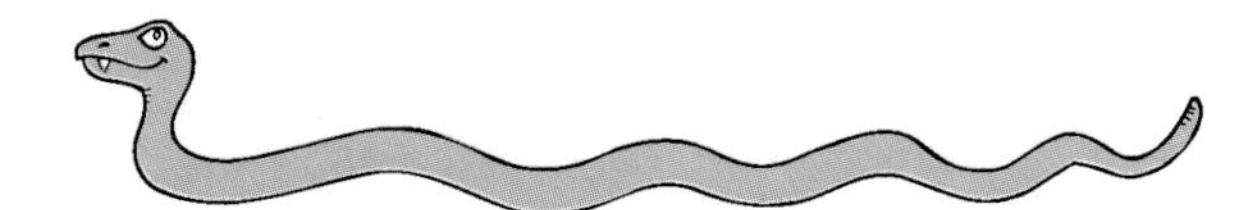

Directions

Draw a snake longer than the longest snake above.

2.

Directions

Draw a snake shorter than the shortest snake above.

3.

Name

Directions

Color the tallest bear BROWN.
Color the shortest bear BLACK.

1.

Directions

Draw a ring around the shortest toy.
Draw an X on the tallest toy.

2.

Ask your child to line up stuffed animals or other toys from shortest to tallest.

Name ________________________________

Directions

Use your bear ruler. Measure how many bears long. Write how many. Draw a box around the longest object. Draw a ring around the shortest object.

1.

________ **bears**

2.

  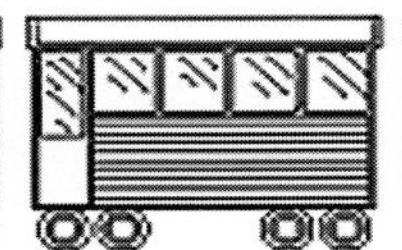 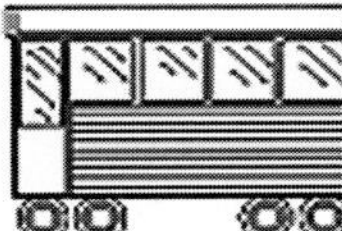 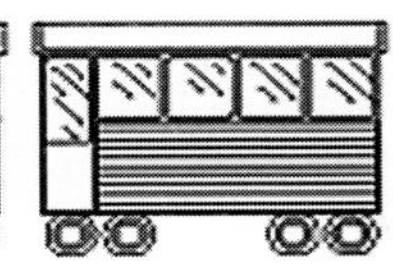 

________ **bears**

3.

4.

________ **bears**

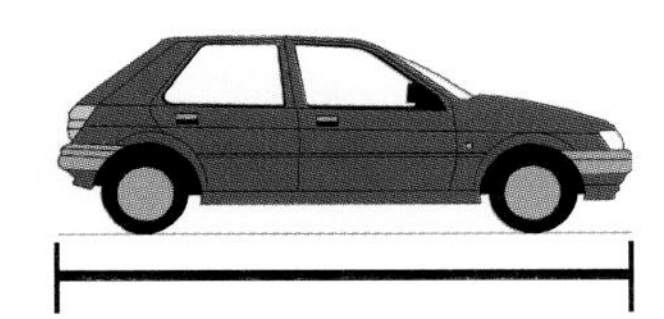

________ **bear**

Name ______________________

Directions

Use your cube ruler. Measure how many cubes high. Write how many. Draw a ring around the shortest animal. Draw a hat on the tallest animal.

1.

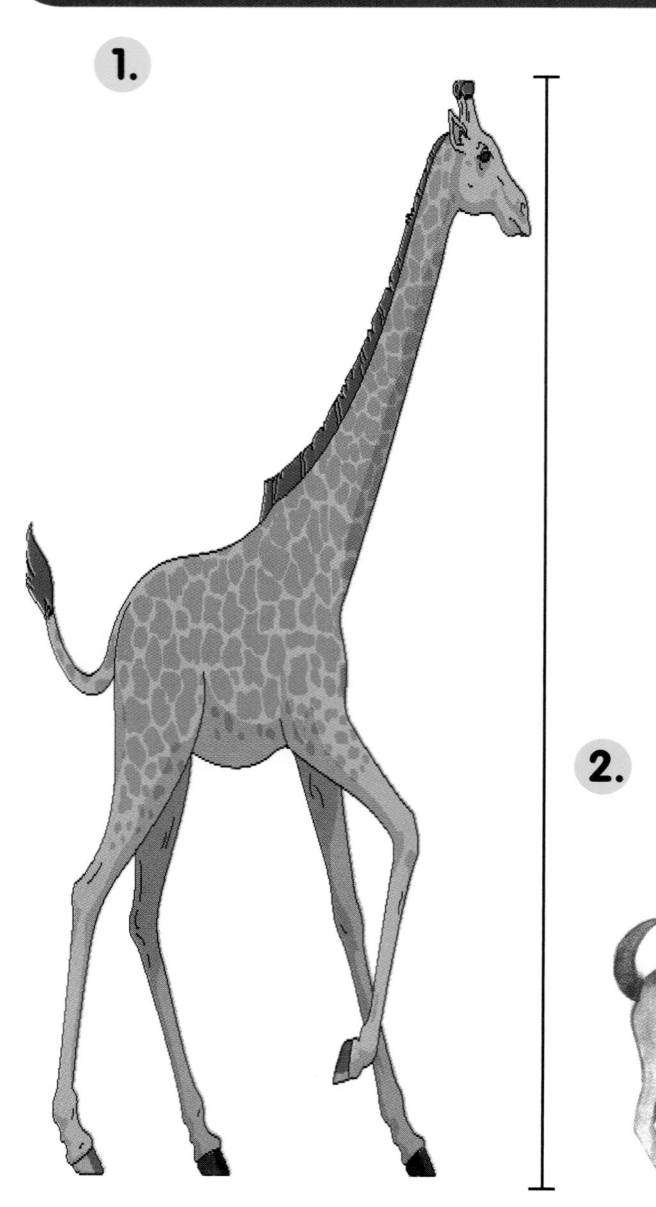

2.

3.

________ cubes ________ cubes ________ cubes

Encourage your child to measure the lengths of various objects using nonstandard units of measure. Ask, for example, How many cars long is this puzzle?

Name ___________________________________

Directions

Color the one that holds more. Use YELLOW
Color the one that holds less. Use GREEN

1.

2.

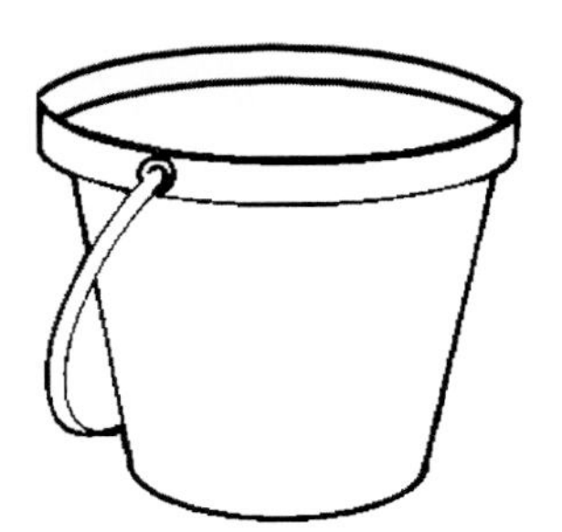

3.

4.

Name ______________________________

Directions

Draw a ring around the object that is heavier.
Write an X on the object that is lighter.

1.

2.

3.

Juice

4.

5.

6.

Play "I Spy" with your child, using heavier than and lighter than as clues.

Name ______________________

Directions

Draw lines. Match animals that are about the same size.

1.

Directions

Write an X on the tallest animal.
Draw a ring around the shortest animal.

2.

Name ______________________________

Directions

Use your cube ruler. Measure how many cubes long.
Fill in the bubble to show about how many cubes.

1.

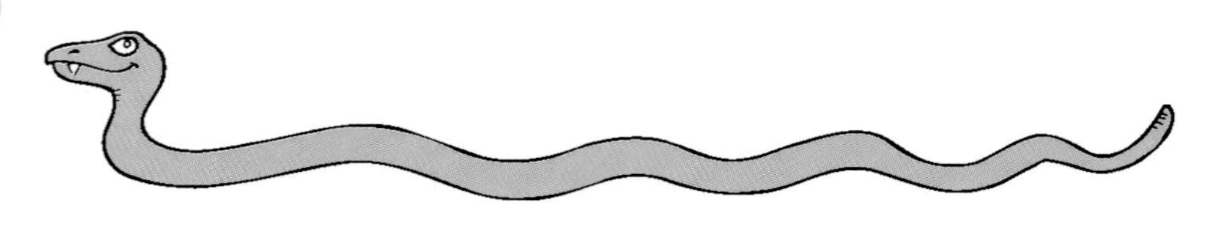

○ 4 cubes

○ 5 cubes

○ 6 cubes

2.

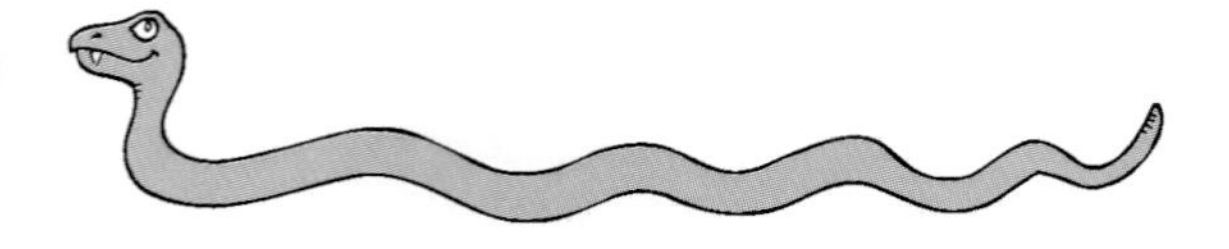

○ 2 cubes

○ 3 cubes

○ 4 cubes

Directions

Which is heavier? Fill in the bubble.

3.

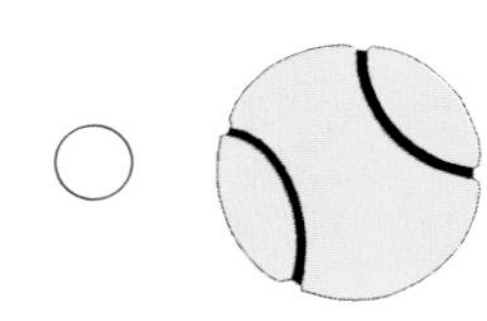

4.

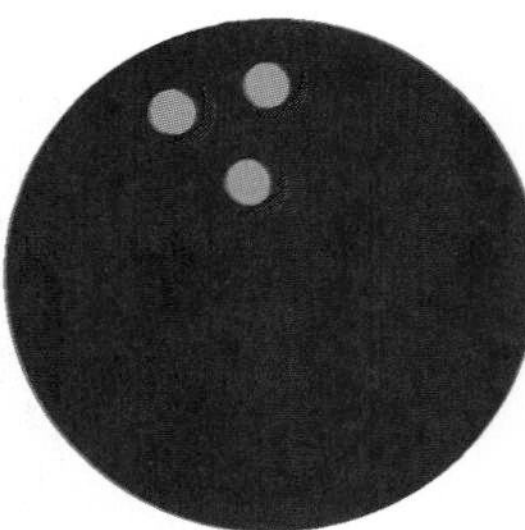

Name ______________________________

Directions

Cut out the shapes below. Work with a partner to make new shapes. How many shapes can you make?

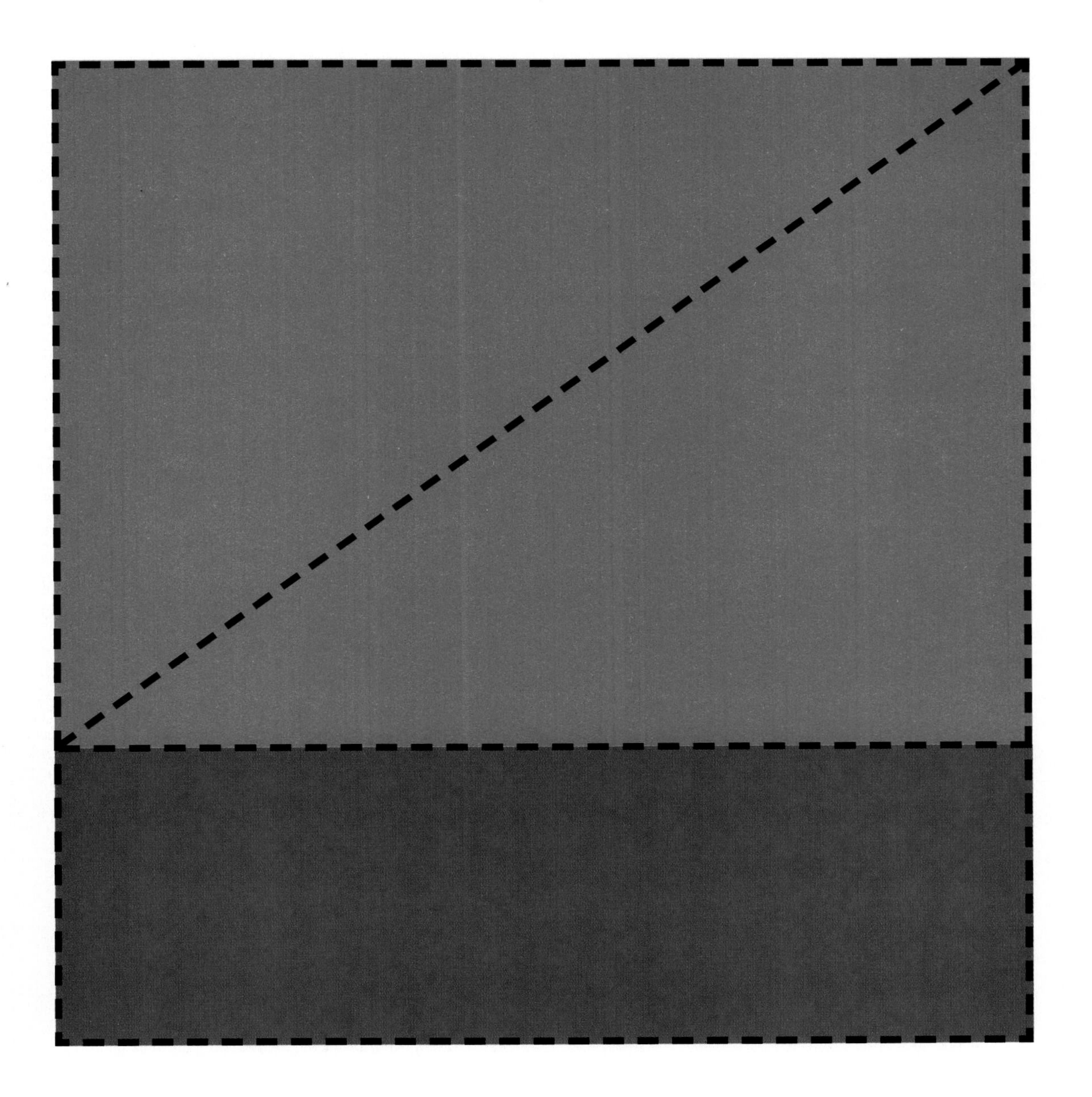

UNIT 4 Vocabulary Words

circle 

sphere 

cylinder

square

cube 

cone 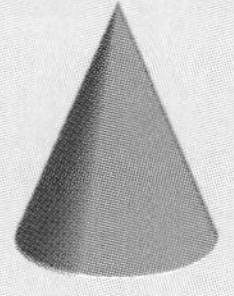

rectangle 

rectangular prism

triangle

A tasty sphere.
12
FOLD
CUT
A sandwich just for me!
5
CUT
Lots of shapes around you and me!
16
FOLD
School Lunch
LUNCH BAG
This book belongs to
1

I look in again.
6
CUT
What else do I see?
11
FOLD
CUT
I open my lunch box.
2
What do you see?
15
FOLD

I take another look.
10
What else do I see?
7
Look at our lunch boxes.
14
What do I see?
3
CUT
CUT
FOLD
FOLD

A cylinder.
8
A cold drink just for me!
9
A rectangular prism.
4
An orange for you and me!
13
FOLD
FOLD
CUT
CUT

Name ______________________________

Directions

Color matching shapes the same color. One shape is done for you.

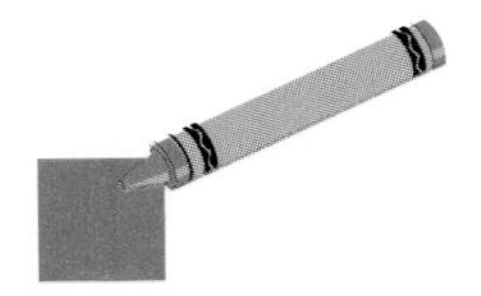
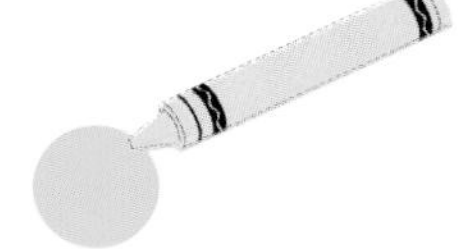

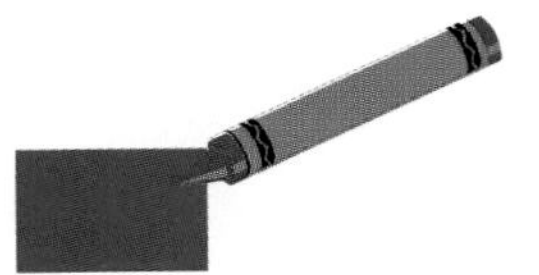

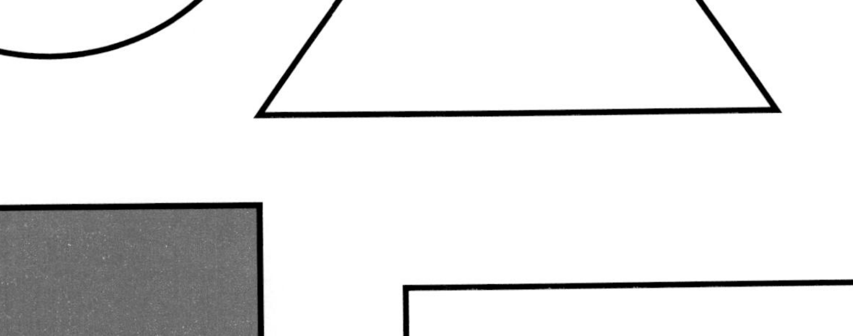
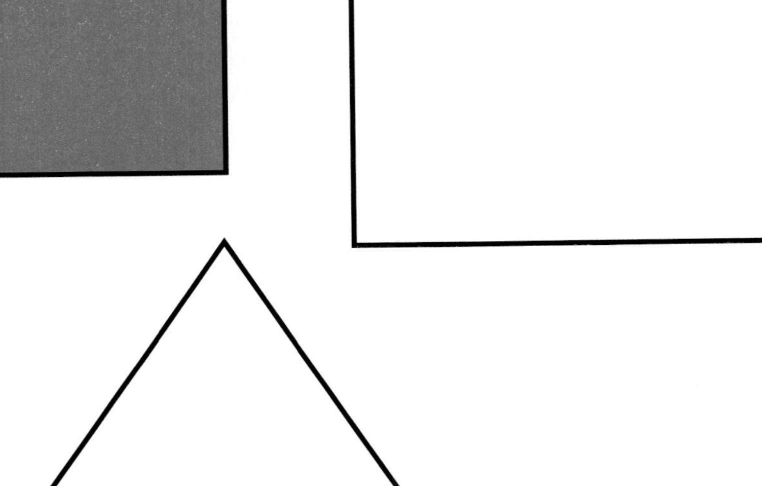

Name ____________________

Directions

Draw a line from the shape to an object that has the same shape. The first one is done for you.

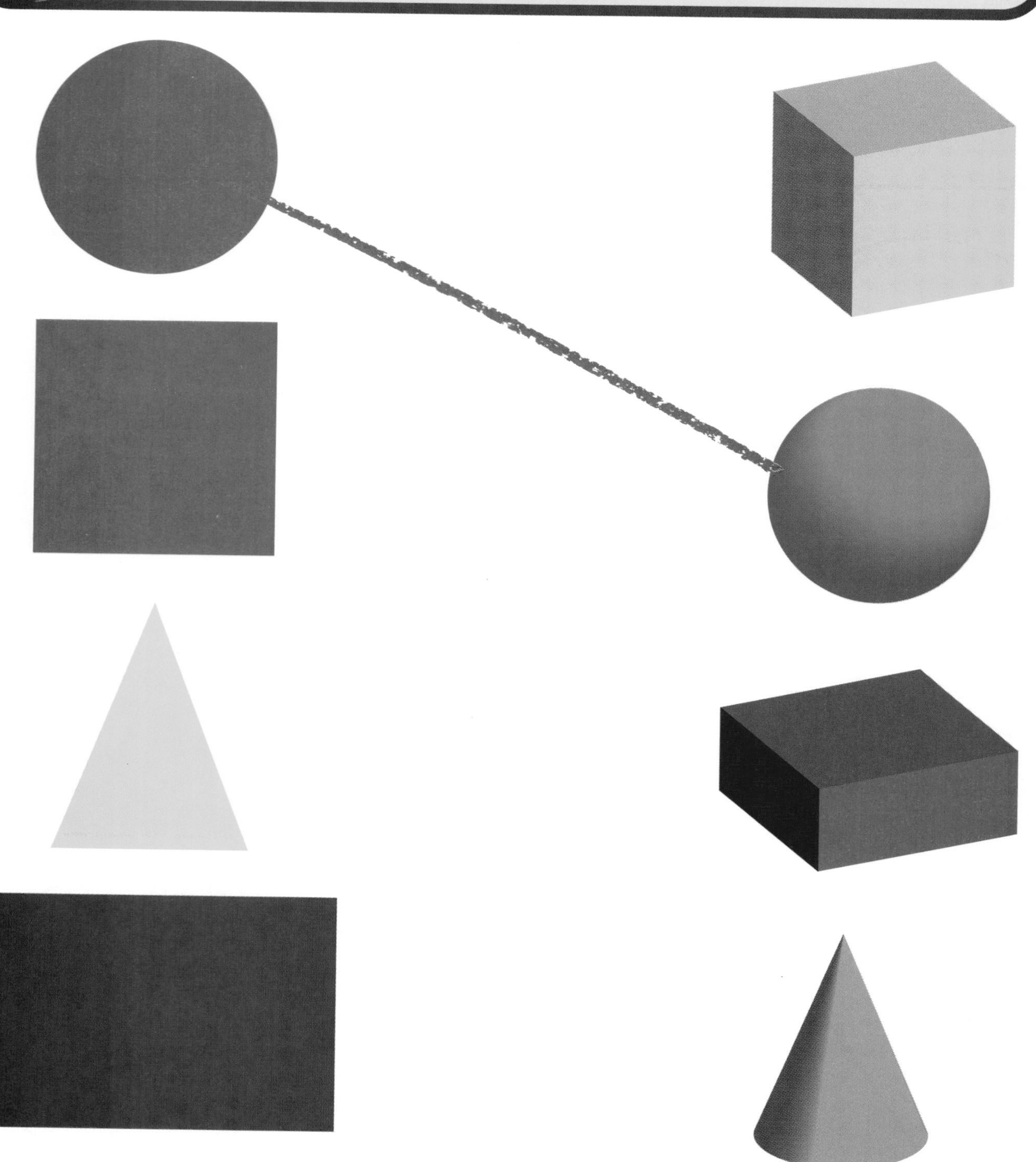

With your child, draw around solid objects, such as cans or boxes, to see the shapes that can be made. Ask your child to explain how the shapes are alike and different.

Name ---------------------------------

Directions

Draw a line from each shape to an object that has the same shape. The first one is done for you.

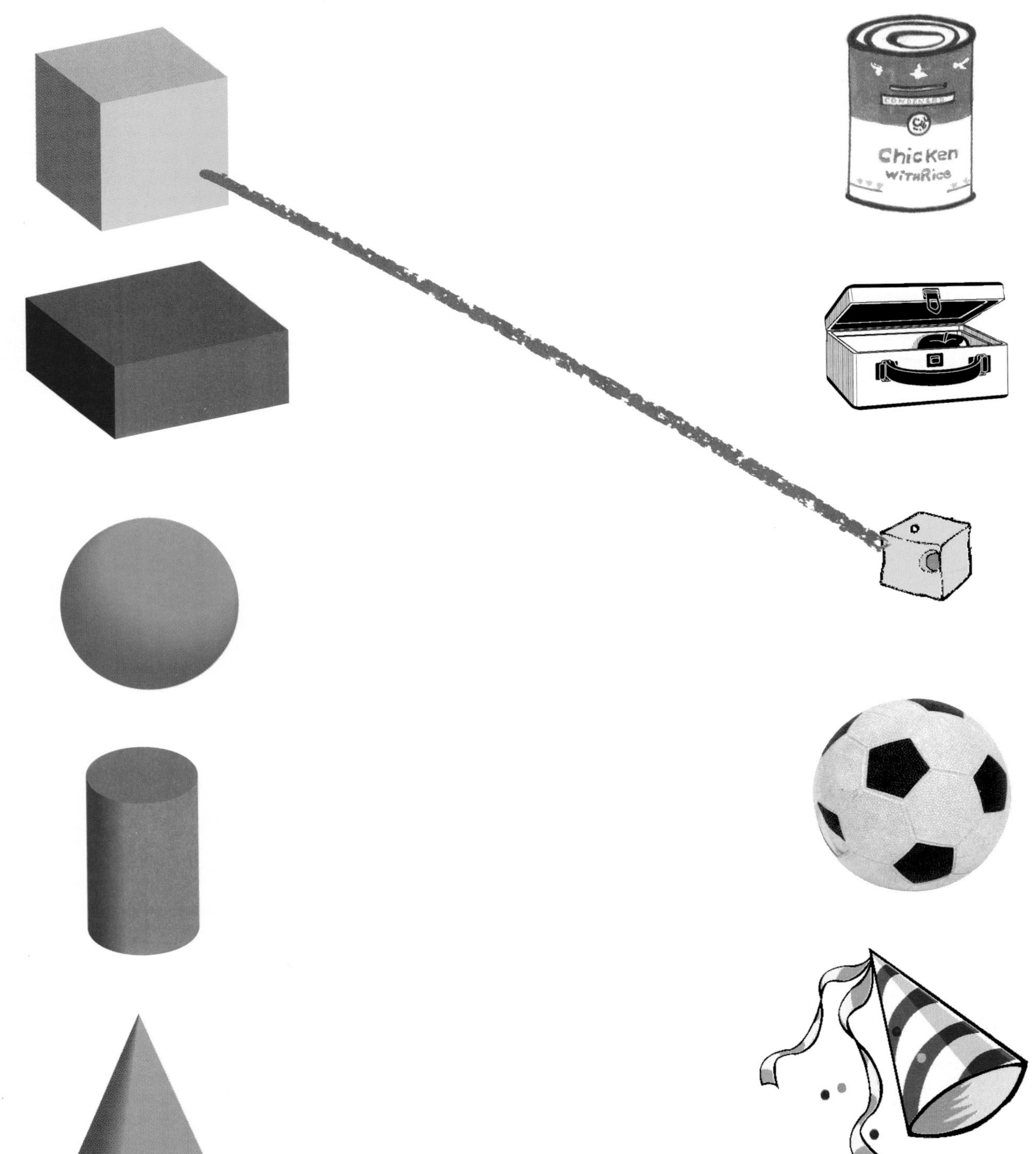

Name ______________________________

Directions

Color the shapes that roll BLUE.
Draw a ring around the shapes that stack RED.
Work with a partner. Talk about the shapes that roll and stack.

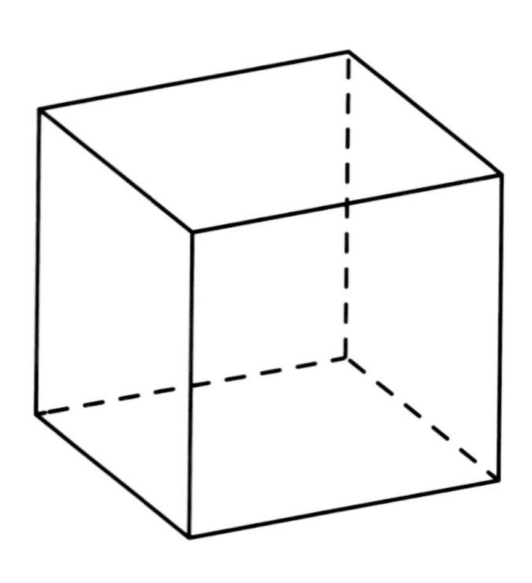

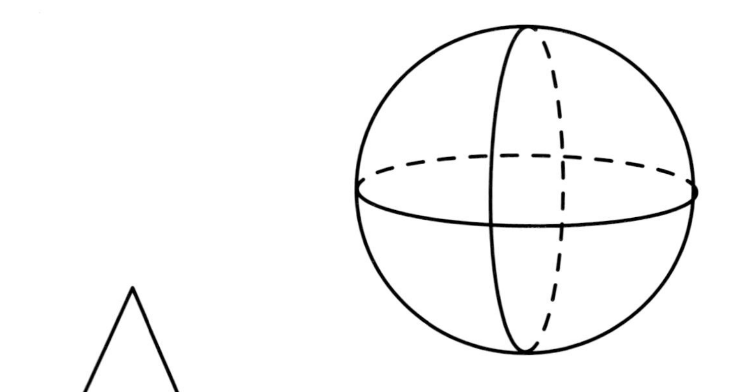

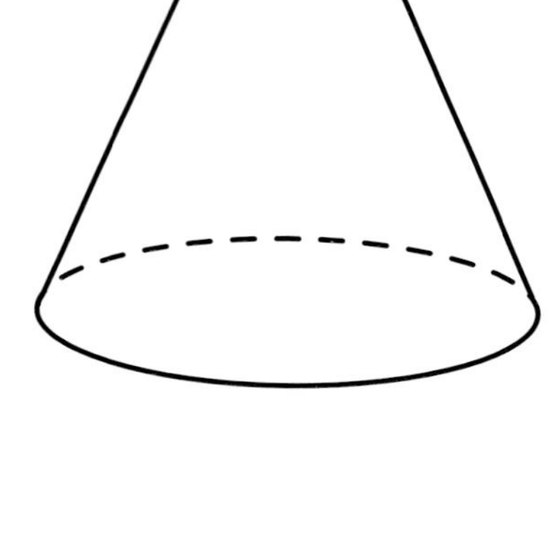

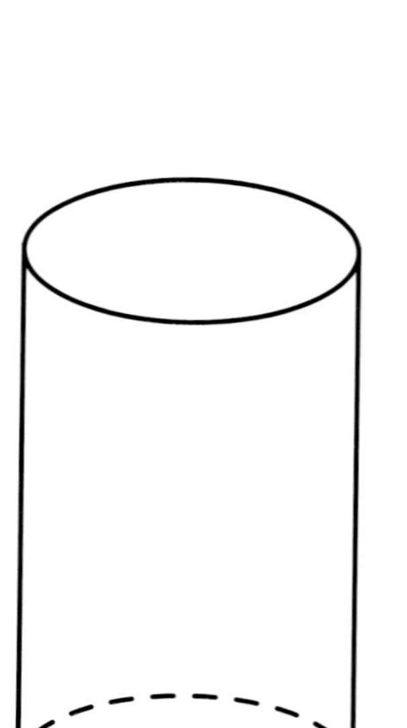

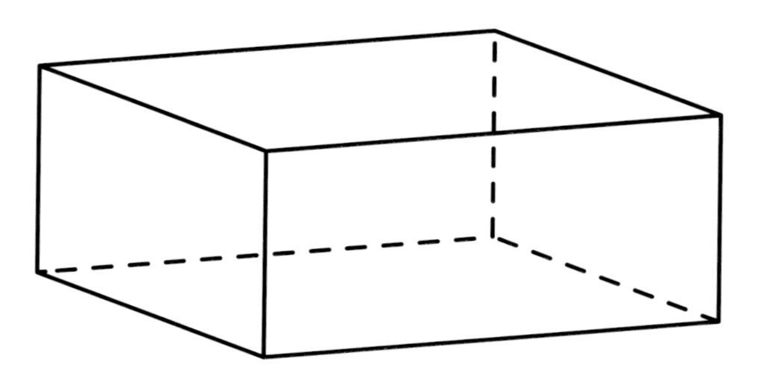

Help your child name shapes of objects in your kitchen.
Name solid shapes and flat shapes.

Name

Directions

Draw lines from the shapes on the left to the matching shapes on the right. The first one is done for you

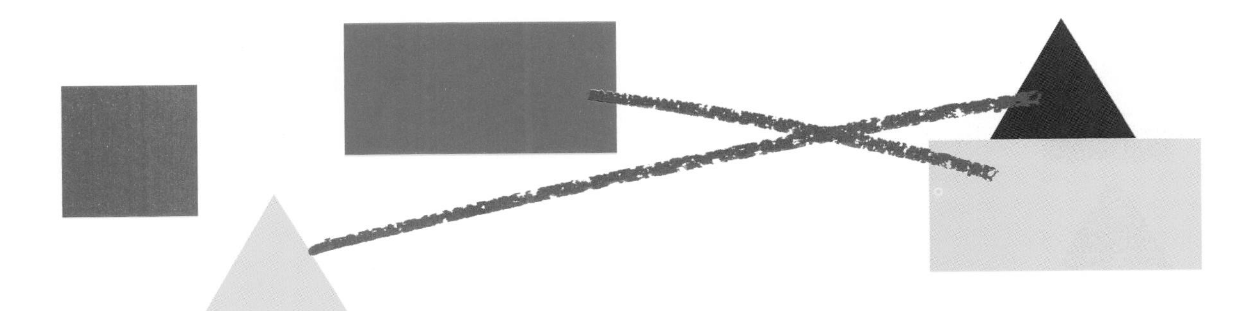

1.

2.

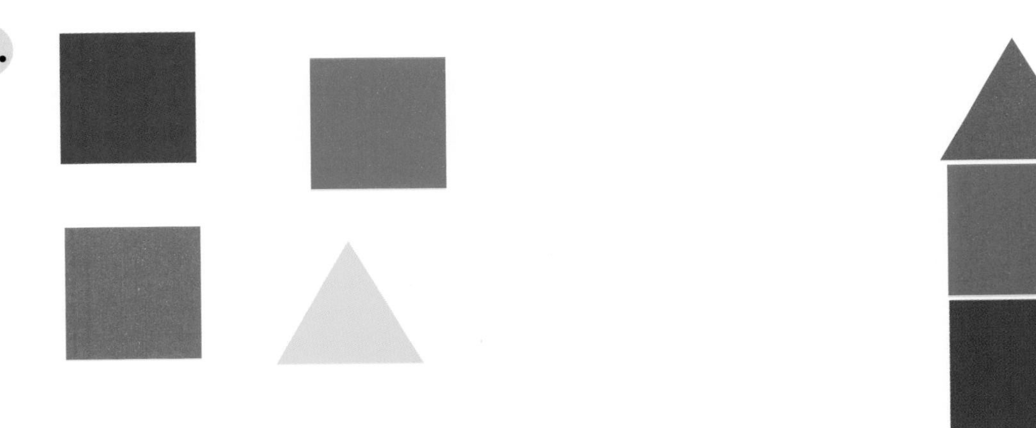

3.

Name

Directions

Look at the shapes on the left. **Draw** a new shape using the shapes on the left. The first one is done for you.

1.

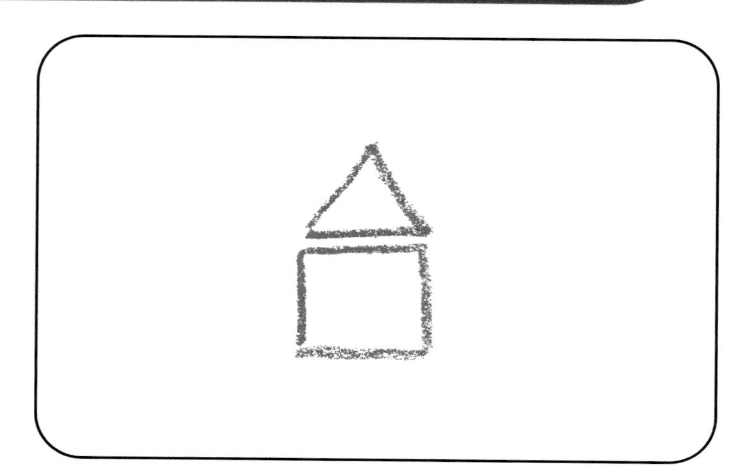

2.

3.

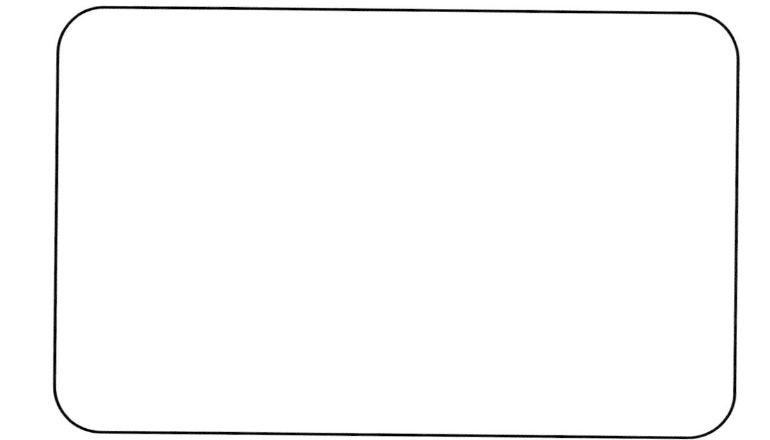

Name ______________________________

Directions

Draw lines to match the shapes you need on the left to make the new shape on the right. The first one is done for you.

1.

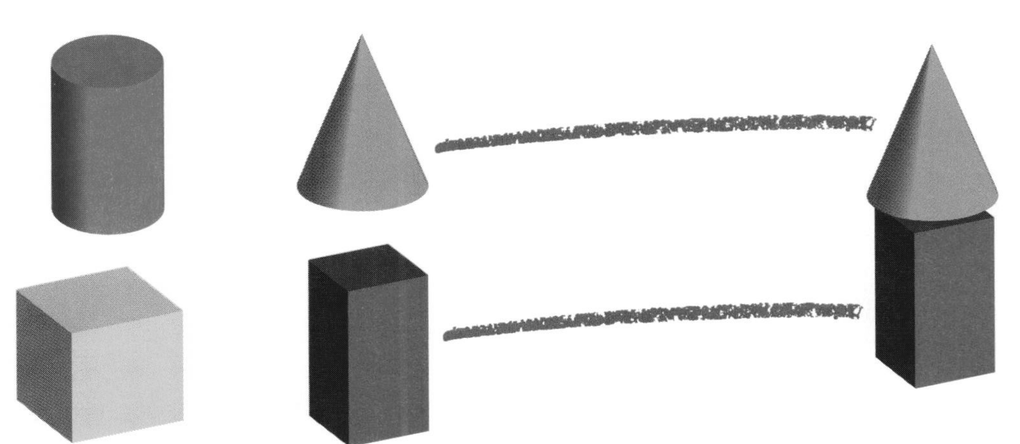

2.

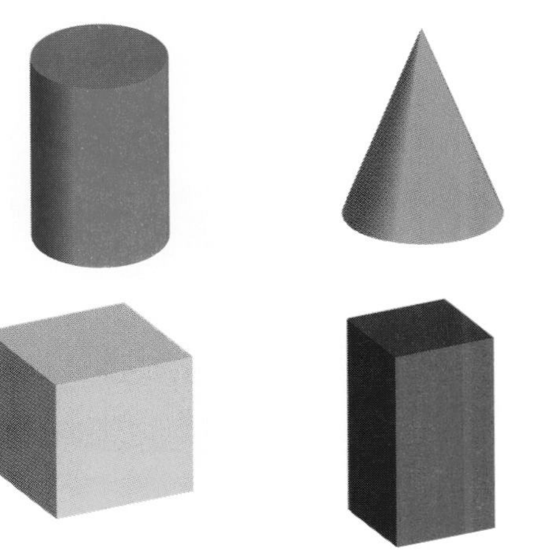

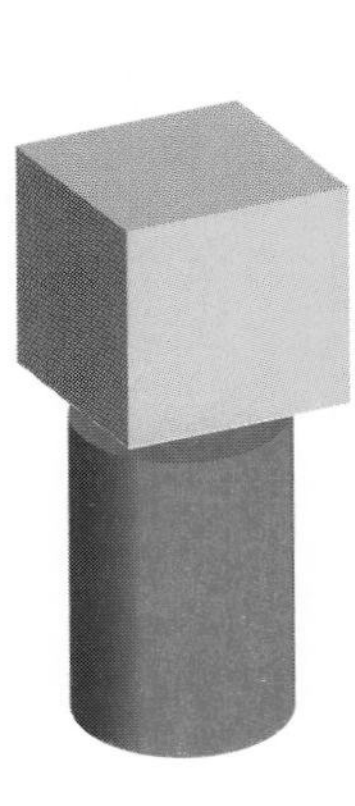

3.

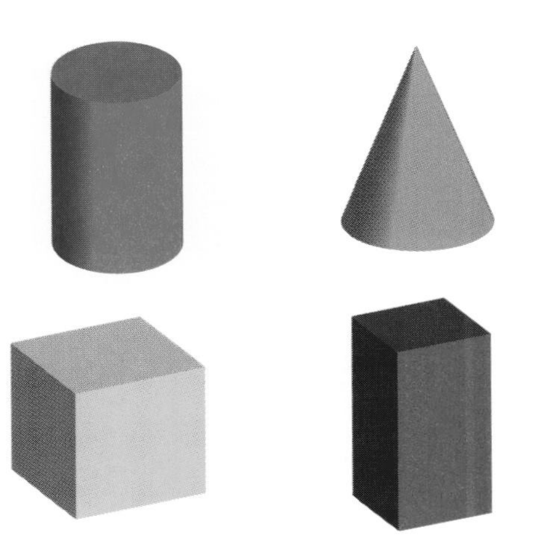

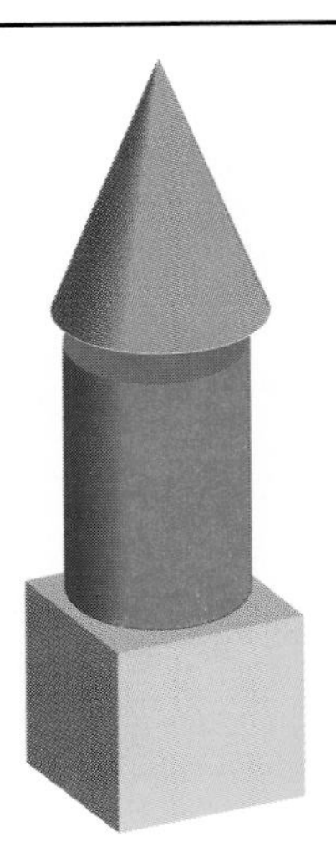

Name ______________________________

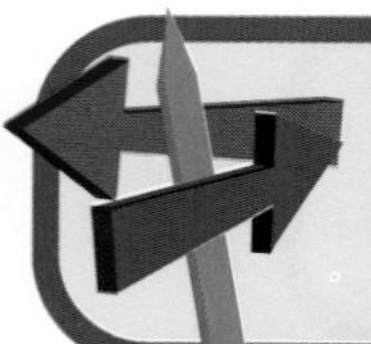

Directions

Can you make the stacks? Draw a ring around the stacks you can build. Talk about why each stack can or cannot be built.

1.

2.

3.

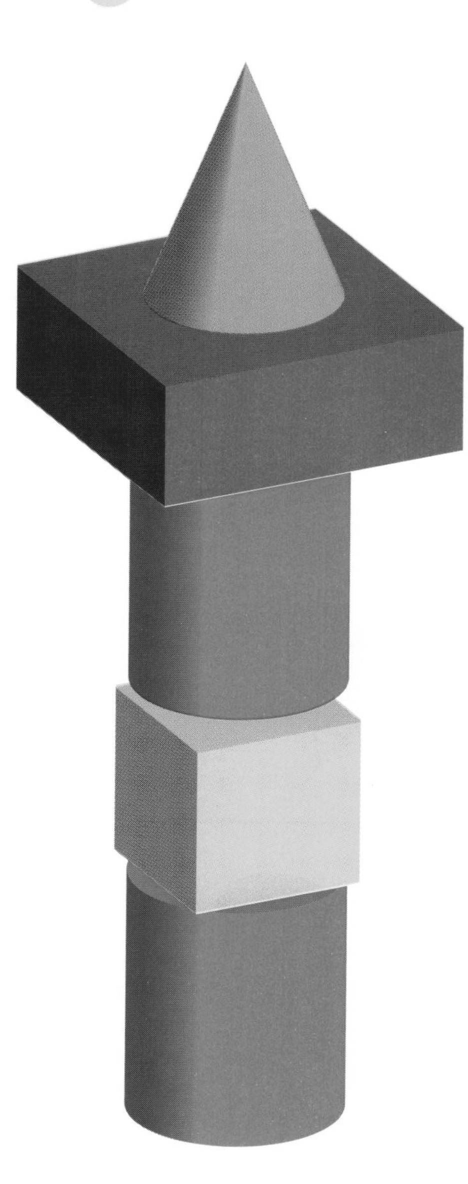

Encourage your child to explore 3-dimensional shapes by building towers and bridges with objects such as blocks, cans, tubes, and boxes.

Name ___________________________

Directions
Look at the shape. Fill in the bubble that shows the same shape.

1.

2.

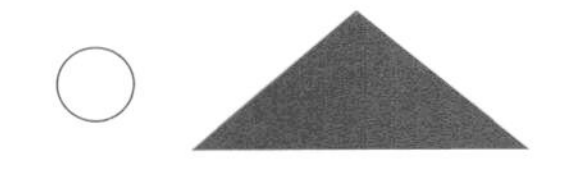

3.

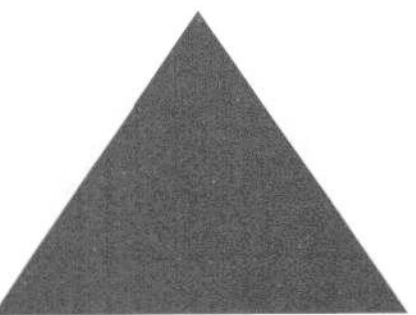

4.

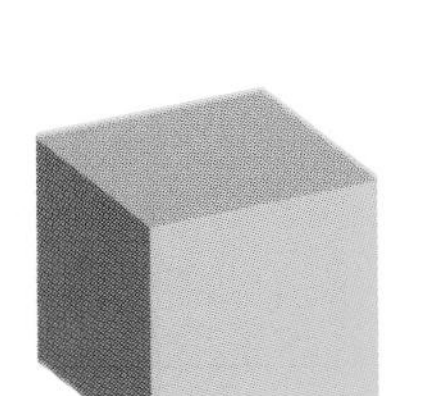

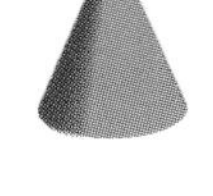

5.

6.

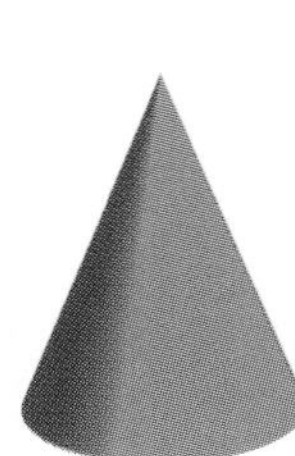

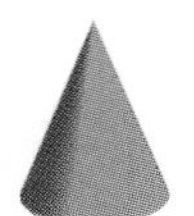

Name ______________________________

Directions

Draw a ring around the new shape you can make with all the shapes on the left.

1.

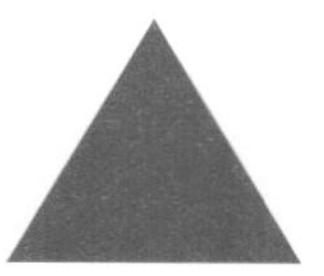

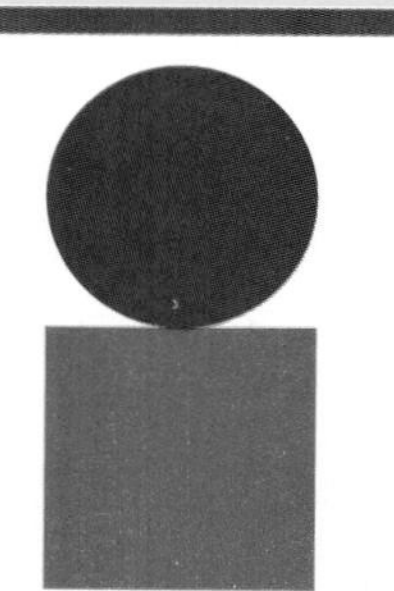

2.

3.

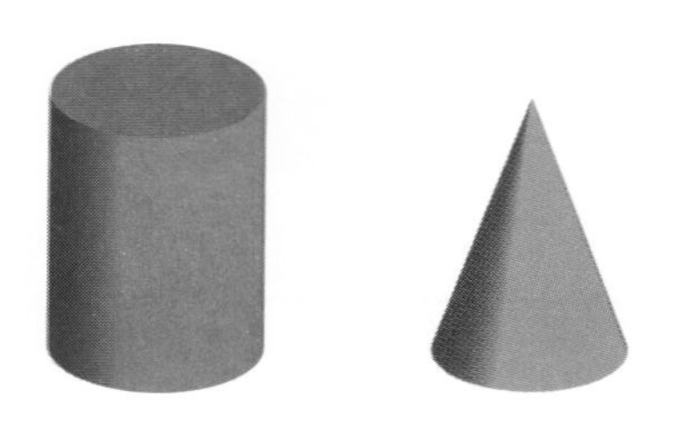

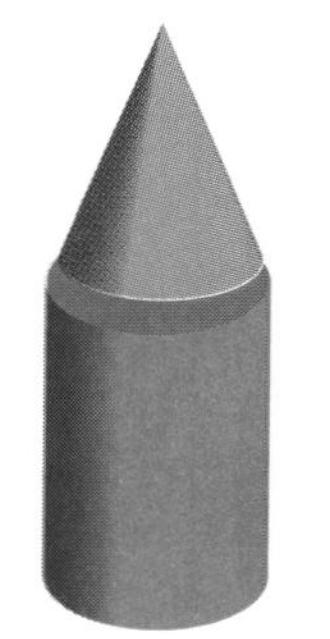
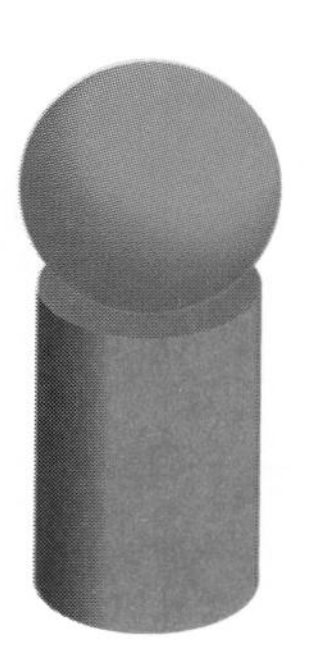
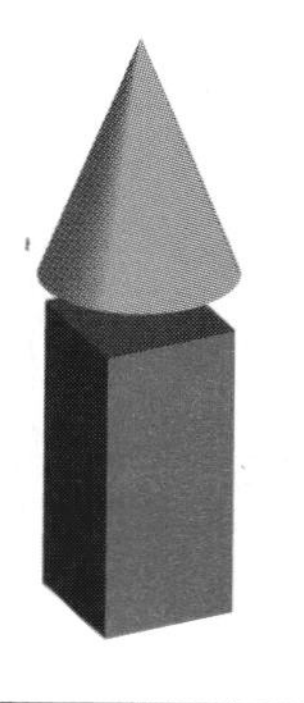

4.

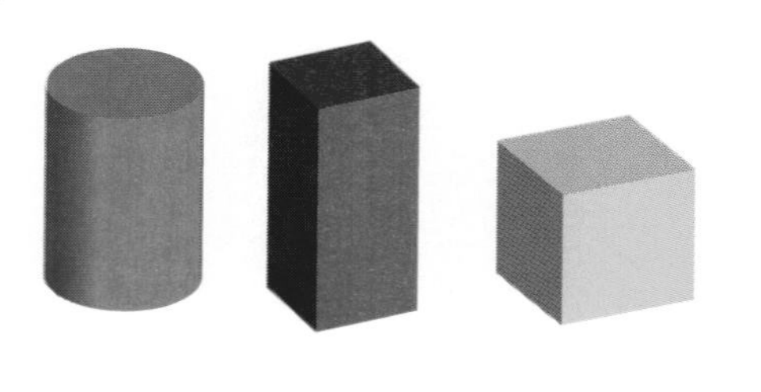

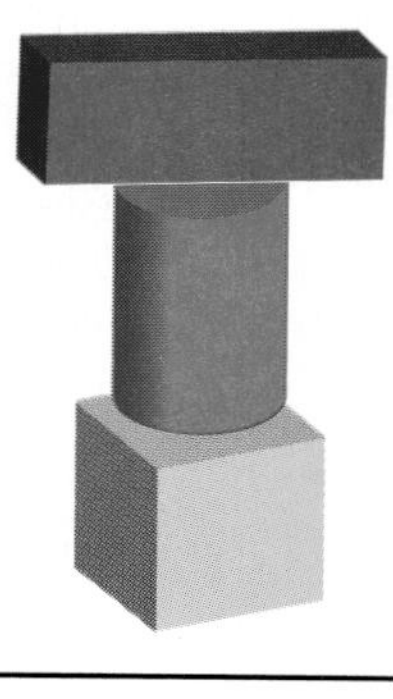

Name ______________________________

Directions

Cut out the popcorn and soda pictures at the bottom of the page. Work with a partner. Look at the first row. Tell what you see. Then place your pictures to copy what you see.

1.

Directions

Look at this row. Tell what you see. Then place your pictures to show what comes next.

2.

UNIT 5 Vocabulary Words

pattern      

next

same

different

EXIT
Two, four, . . .
FOLD
12
CUT
Six of us.
5
CUT
Do you think you could hear them shout?
FOLD
16
On the Ride
This book belongs to
1

Eight of us.
6
CUT
How many again?
FOLD
11
Who wants to go for a ride?
2
FOLD
We saw them come out.
15

What fun! What fun!
10
CUT
FOLD
Ten of us.
7
CUT
We saw them go in.
14
Two of us.
3

Up, up, up, we go!
8
CUT
Whee! Now, down we go!
9
FOLD
CUT
Four of us.
4
FOLD
. . . six, eight, ten.
13

Name ______________________________

Directions

Look at the pattern in the top row. Tell what you see.
Use RED and YELLOW to copy the pattern in the next row.

1.

     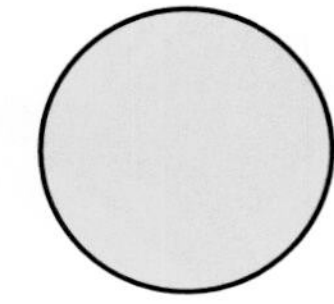

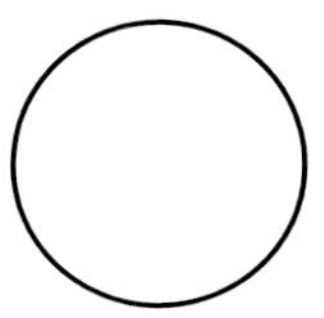 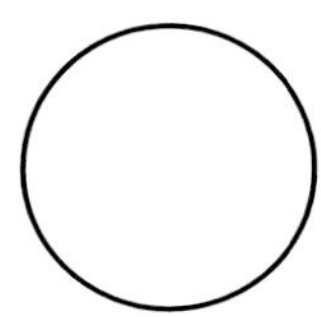 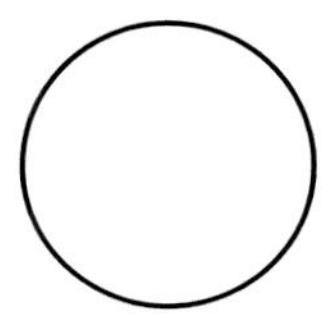 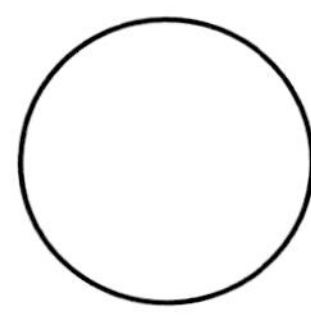 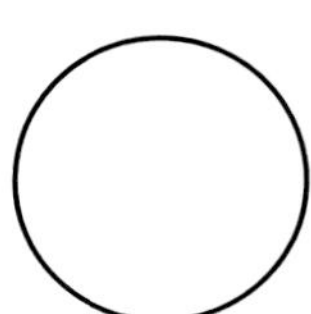 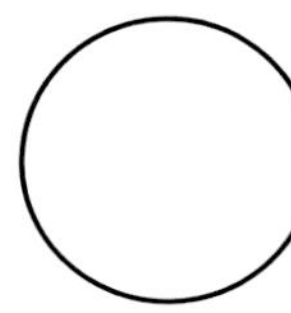

2.

3.

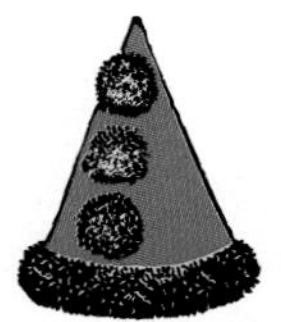 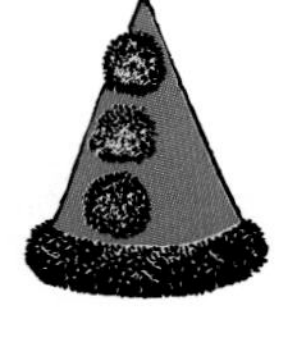    

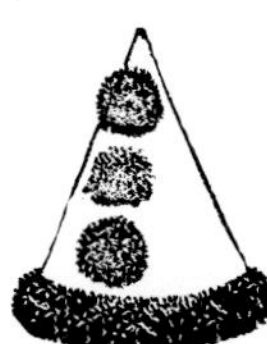 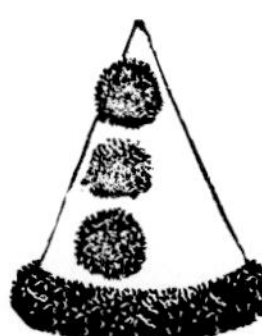 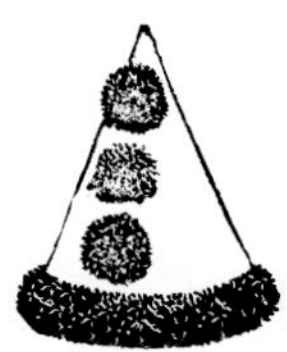 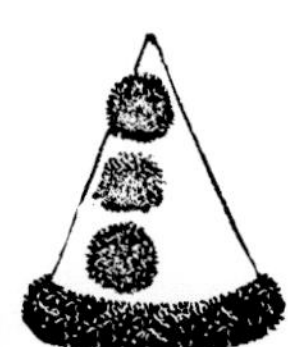 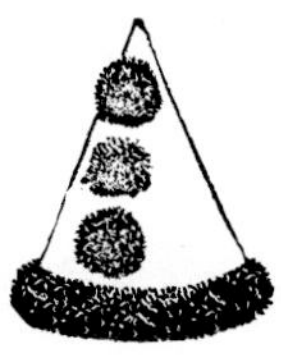 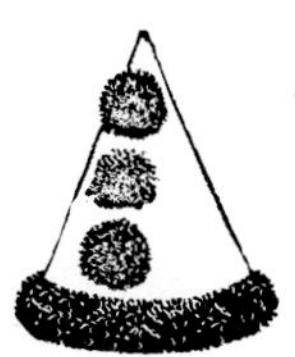

Name

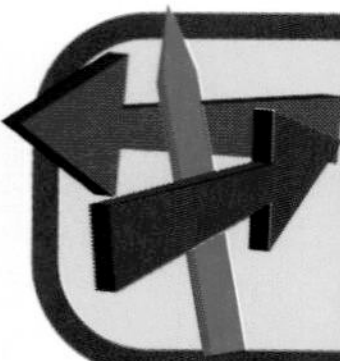

Directions

What is missing? Draw a line to the picture that completes the pattern. The first one is done for you.

1.

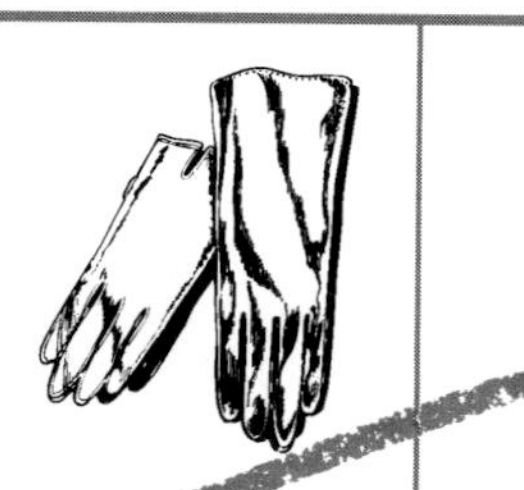

2.

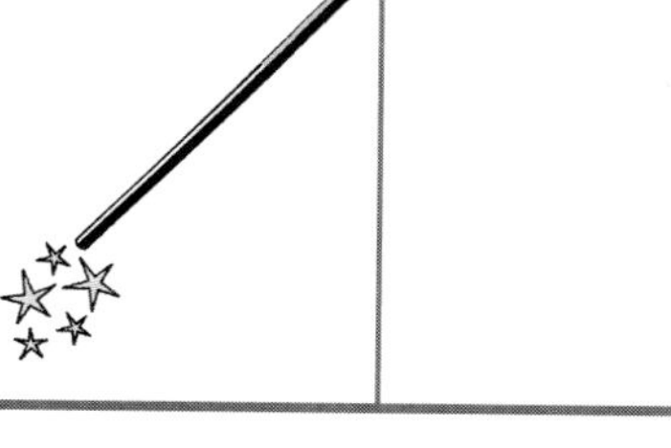

3.

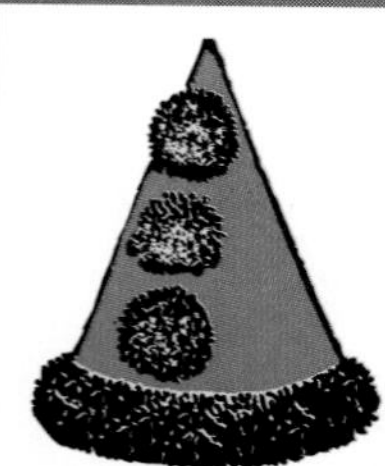

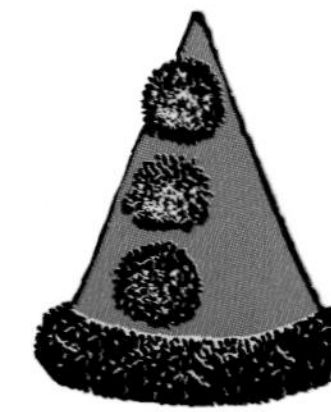

Gather together different sizes and colors of socks. Use them to create different patterns with your child.

Name ______________________________

Directions

Look at the pattern. Tell what you see. Draw a ring around what comes next.

1.

2.

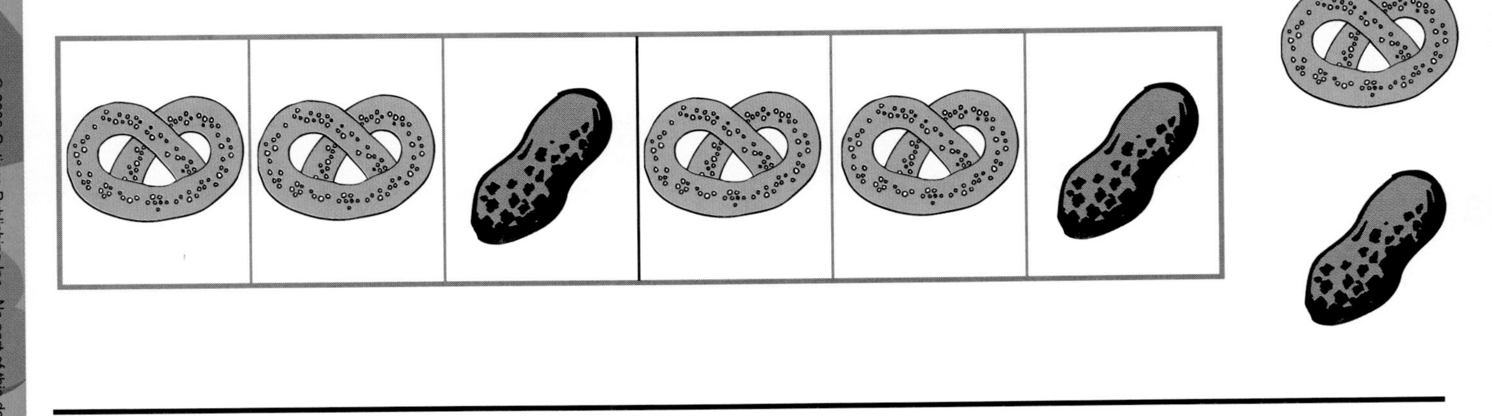

3.

Name ______________________________

Directions

Look at the pattern. Tell what you see.
Color what comes next. Use  .

1.

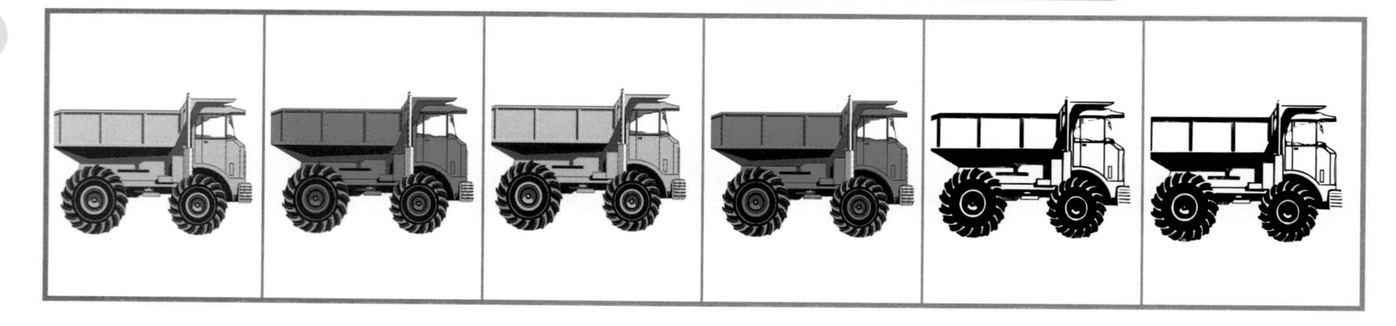

2.

3.

4.

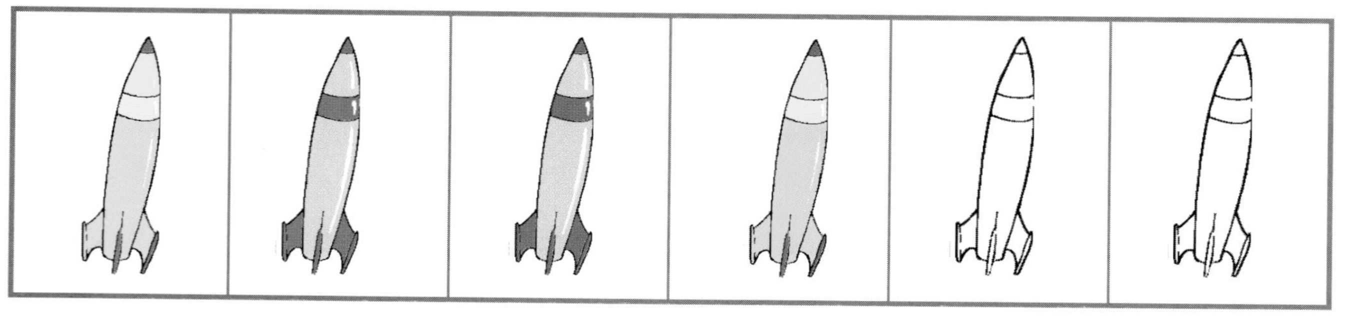

Gather together three kinds of common household objects, such as forks, spoons, and knives. Arrange them in simple patterns, at first using only two kinds of utensils. Ask your child what the utensil comes next each time. Then make another pattern using all three kinds of utensils

Name

Directions

We find patterns in numbers, too! The elephant jumps on each stand. Trace the elephant's jumps and count each number.

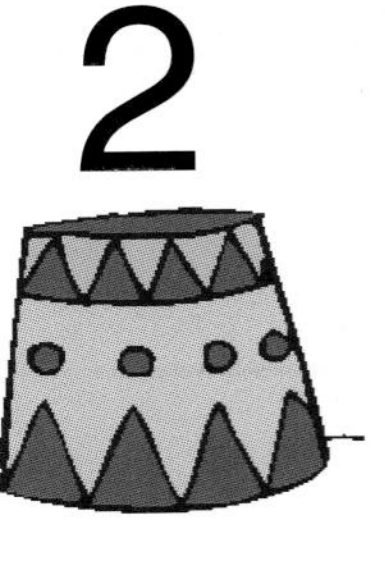

What would the next number be? ______________

Tell how you know.

Name

Directions

Look at the pattern. Tell what you see. Draw the shapes to show what comes next. Draw as many shapes as you can for each necklace.

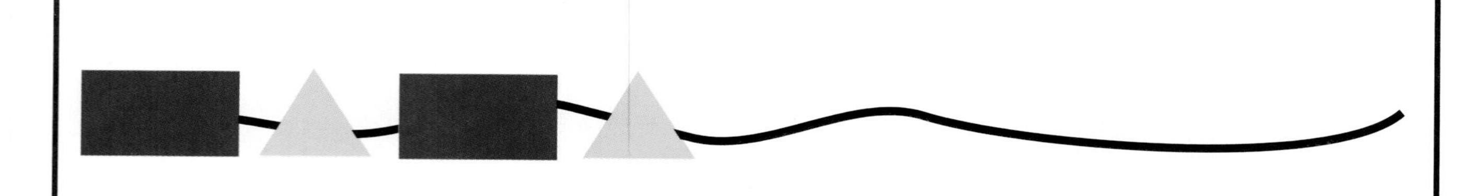

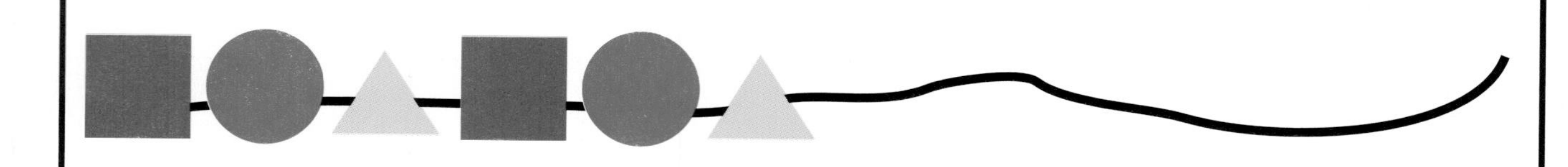

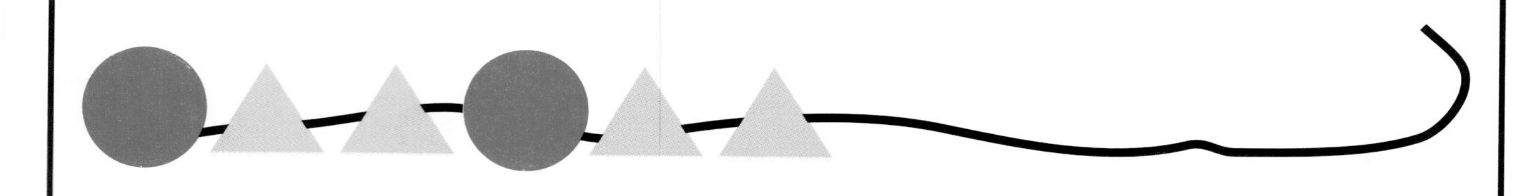

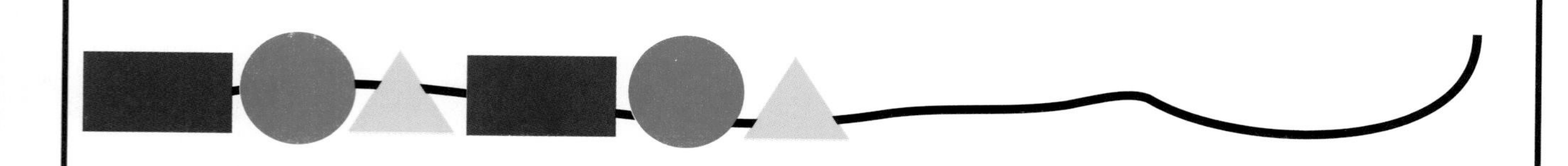

Give your child colored pasta and string to make a pattern necklace or bracelet. Ask your child to describe the pattern.

Name ________________________________

Directions

Color one more square for each row. Look at the pattern. Say the numbers in order.

1. black	■							
2. yellow	■	■						
3. brown	■	■	■					
4. orange								
5. blue								
6. red								
7. green								
8. purple								

Name ___________________________

Directions

Find the clowns with no dots. Draw the correct number of dots to complete the pattern.

1.

1

2

3

4

5

6

7

8

9

10

Directions

What is missing? Write the number in each box.

2.

1	2	3	4		6	7	8

3.

3	4	5		7		9	10

Give your child paper clips to make chains. Count the number of paper clips in the chain. Ask your child to add one more paper clip and count the paper clips again.

Name

Directions

Fill in the bubble to show what comes next.

1.

2.

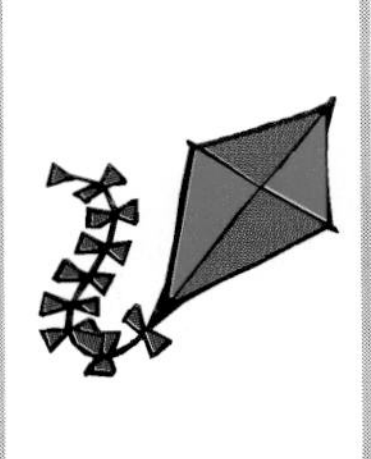

Directions

What is missing in the pattern? Fill in the bubble.

3.

4.

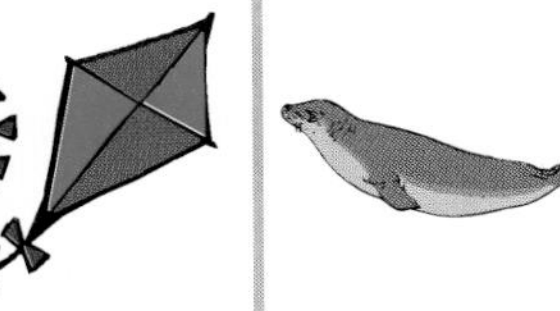

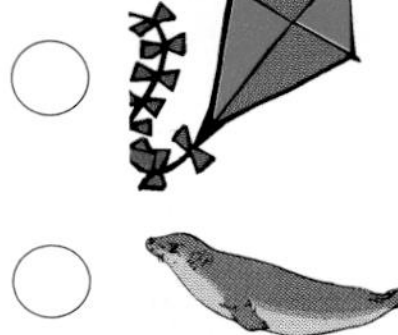

Name ______________________________

Directions

Look at the pattern. Draw the shapes that come next.
Draw as many as you can.

1.

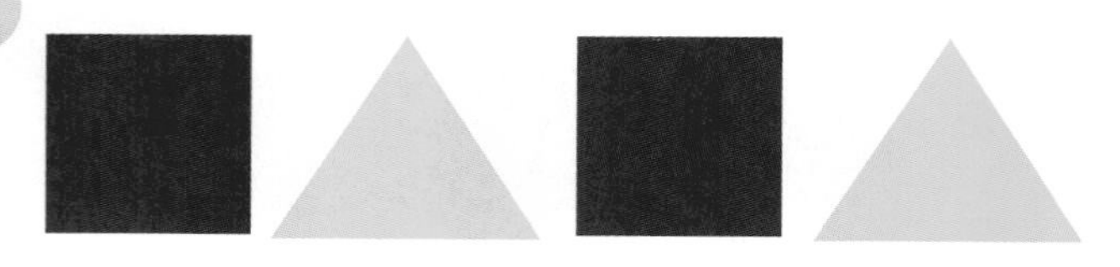

2.

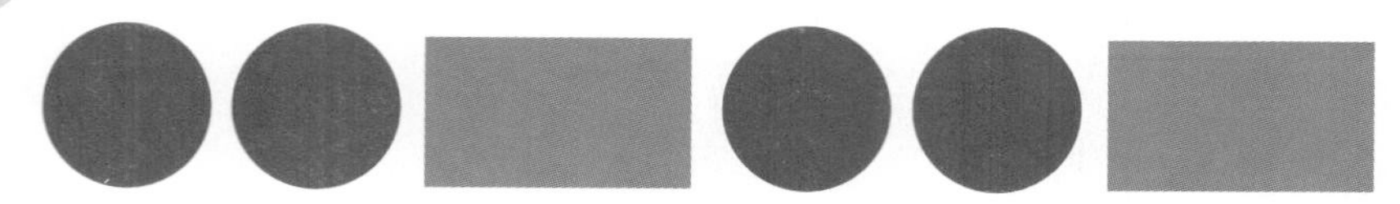

Directions

Draw 1 more scoop of ice cream on each cone.

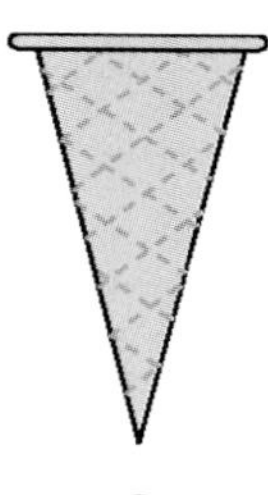
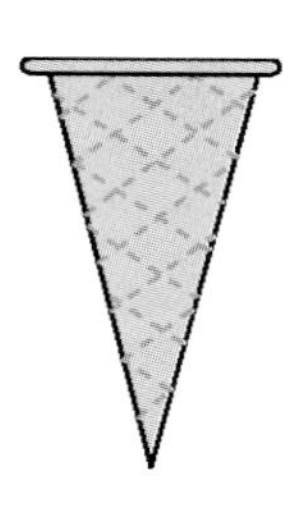
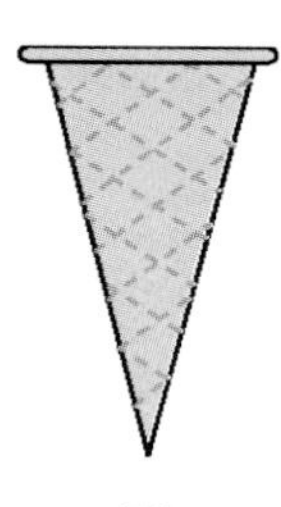

1 2 3 4 5

Name ______________________________

Directions

Cut out the cars and trucks. Work with a partner. Take turns. Match the colors of the cars on the graph. Have your partner tell how many of each color.

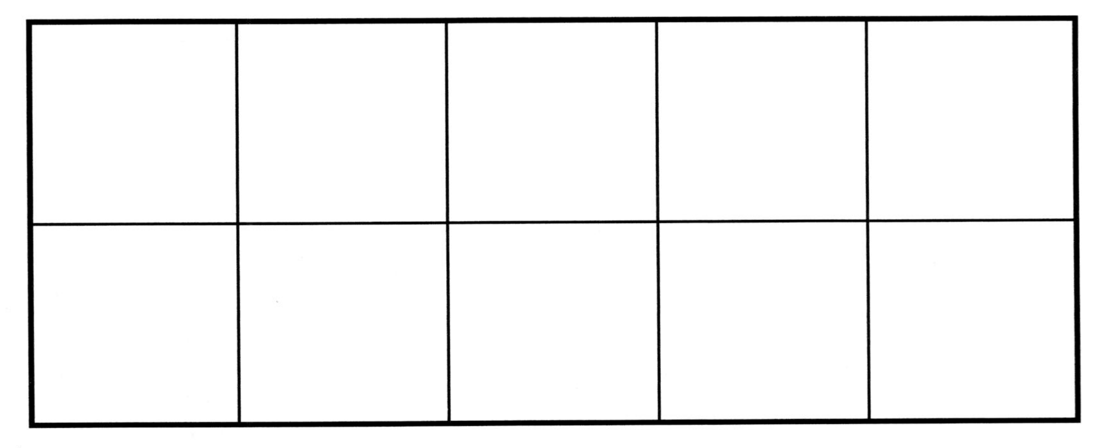

UNIT 6 Vocabulary Words

same

more

most

bar graph

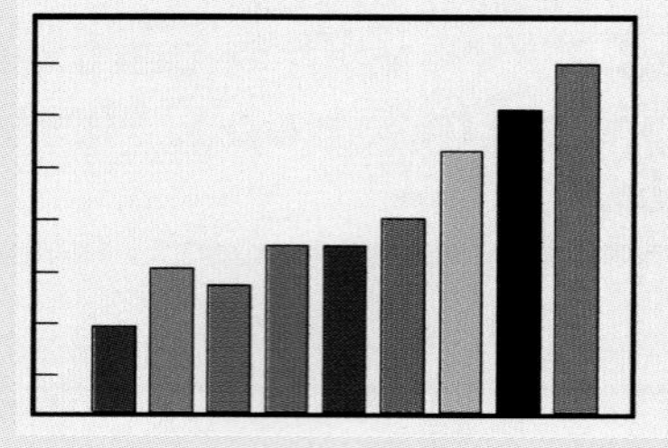

picture graph

Blue	
Striped	
Orange	

How many vans did we see?
12
FOLD
VANS
We counted vans.
5
CUT
CUT
Wait! Did we count these?
FOLD
16
Counting Cars, Trucks, and Vans
10TH
This book belongs to
1

SMALL TRUCKS
We counted small trucks.
6
Traffic in front of school
FOLD
We made a picture graph.
11
CUT
Mr. Hill always says to look both ways.
SCHOOL
2
CUT
Traffic in front of school
FOLD
What did we see the least of?
15

We put all the
pictures together.
10
FOLD
BIG
TRUCKS
II
We counted big trucks.
7
CUT
CUT
Traffic in front of school
What did we see the most of?
14
FOLD
Today we did.
3

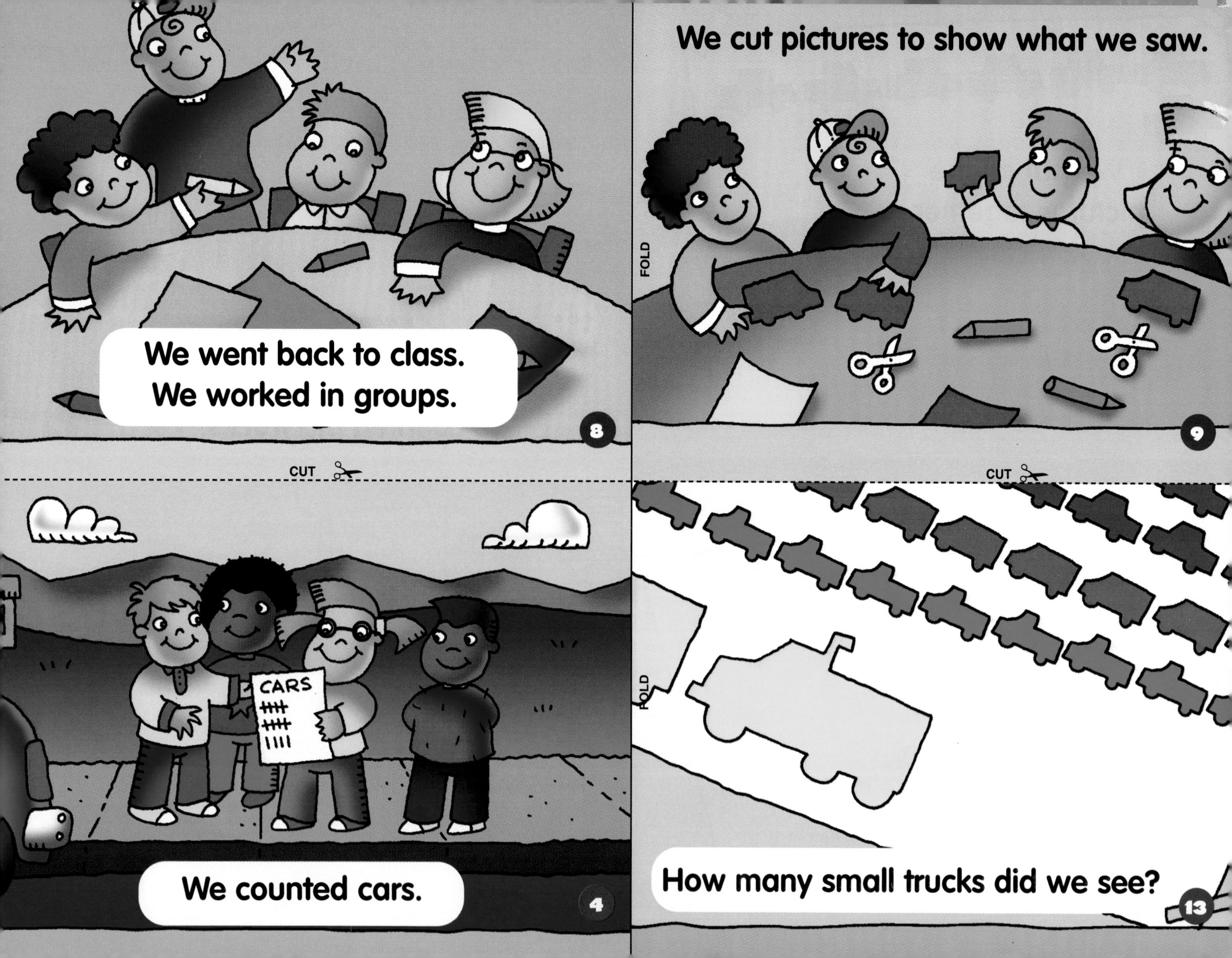
We went back to class.
We worked in groups.
8
We cut pictures to show what we saw.
FOLD
9
CUT
CUT
CARS
We counted cars.
4
FOLD
How many small trucks did we see?
13

Name

Directions

Cut out the pictures. Make a graph. Paste the pictures on the graph. Tell how many of each.

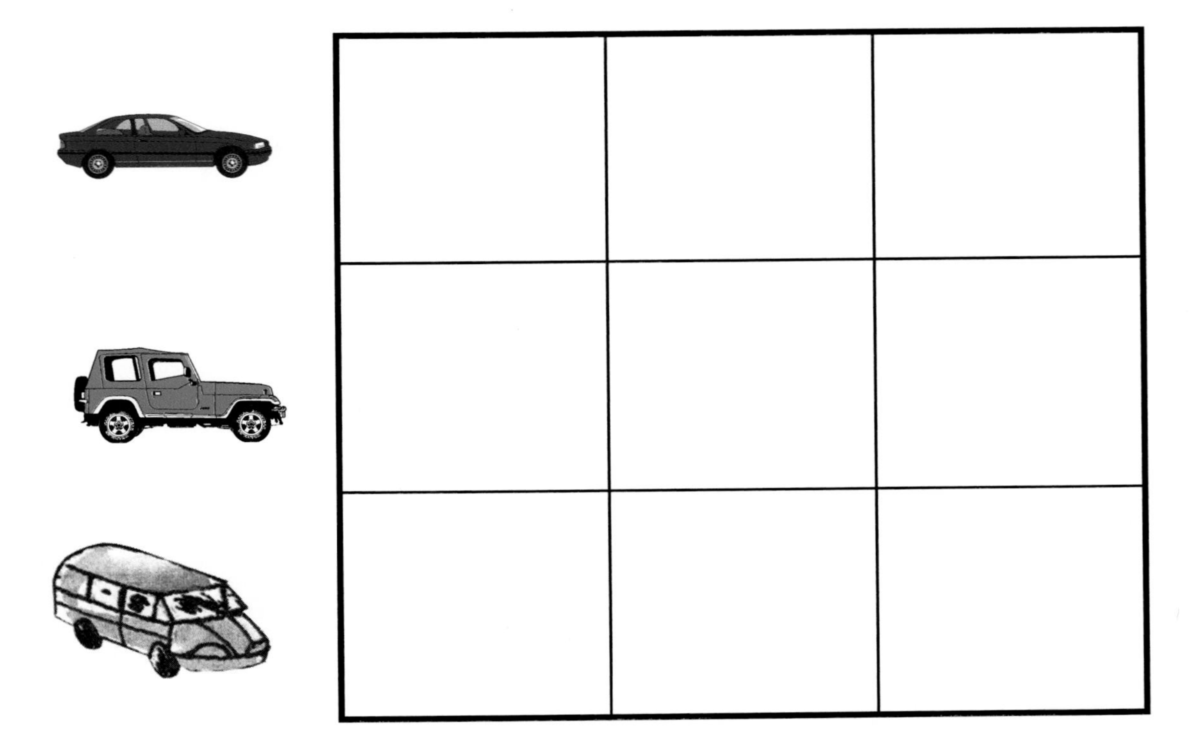

Name ________________________________

Use RED to draw lines to put cars in the car garage.
Use GREEN to draw lines to put trucks in truck garage.

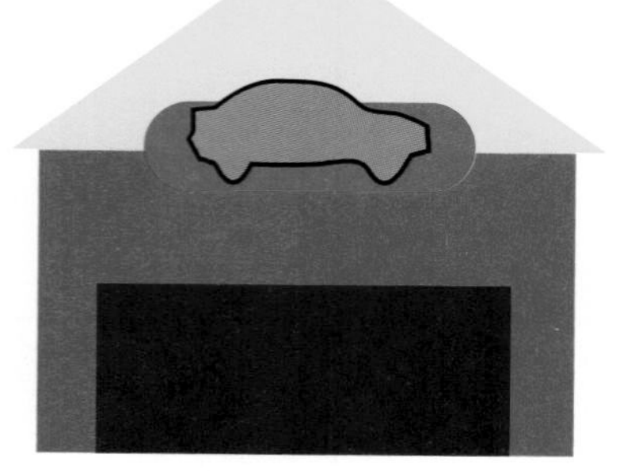

Show your child a collection of socks in two or three colors. Ask your child to sort the socks by color.

Name

Directions

Use BLUE to graph the cars.
Use YELLOW to graph the buses.
Draw a ring around the answer.

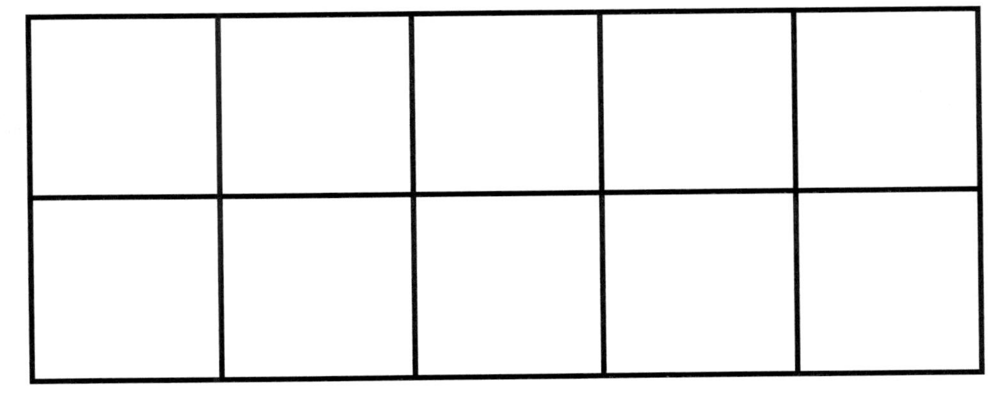

1. How many  ? 2 3 4 5

2. How many 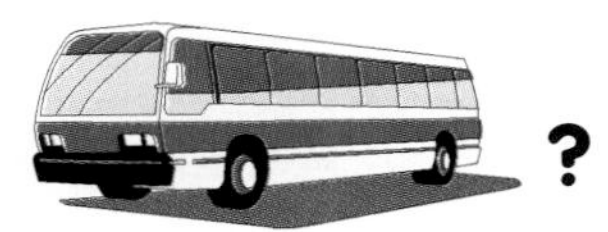 ? 2 3 4 5

3. Which has more?  

Name

Directions

Use PURPLE to graph the .
Use GREEN to graph the .
Use ORANGE to graph the .
Draw a ring around the answer.

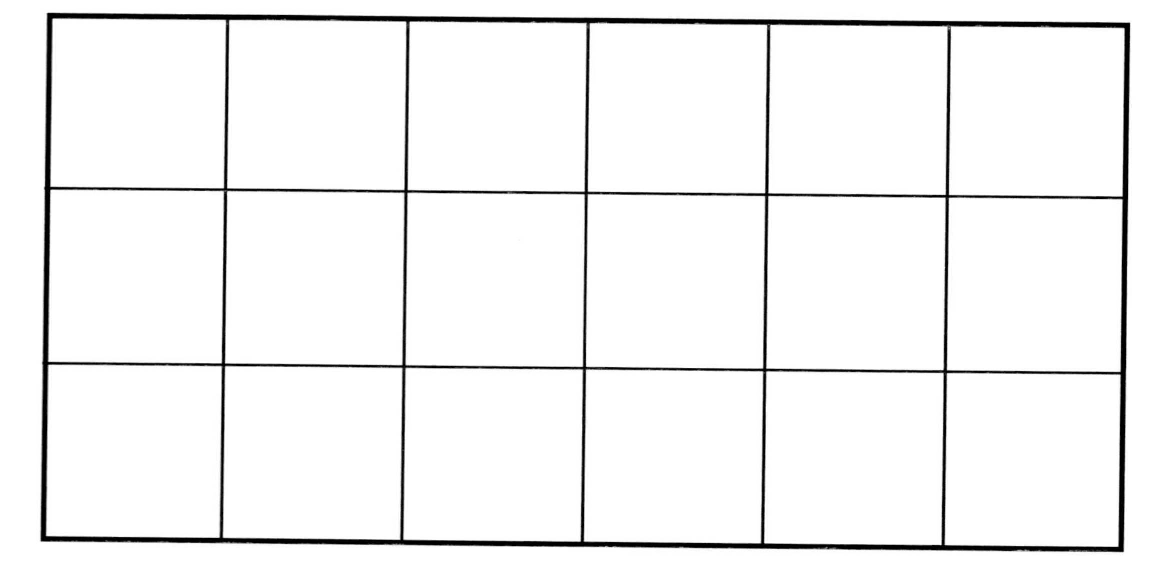

1. How many  ? 1 2 3 4 5

2. Which has the most?   

Show two groups of objects to your child. Give your child some counters such as cubes or chips. Have your child show a corresponding number of counters for the objects in each group.

Name ______________________________

Directions

Adam and Dan have toy race cars. They make a picture graph. Use the graph to answer the questions.

				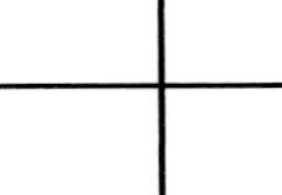		

1. How many   ? 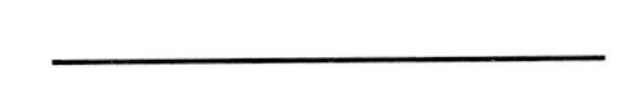 ______________

2. How many  ? ______________

3. How many  ? ______________

4. How many in all? ______________

Name ________________________________

Directions

Count the cars. Color a car on the graph for each car. The first line of the graph is done for you. Then use the picture graph to answer the question.

Color	How Many?

1. How many  ? ______________

Make two groups of six things. Ask your child to show a group with more and a group with fewer things. Have your child tell how many are in each group.

Name ______________________________

Directions

Sara made a graph. She colored a box for each truck. Use the graph to answer the questions.

1. How many  ? ____________

2. How many  ? ____________

3. How many in all? ____________

4. Which has more?  

5. Sara saw 1 more  .

   Use to show it on the graph above

Name ________________________________

Directions

Read the bar graph. Use the graph to answer the questions.

1. **How many**  **?** ______________

2. **How many**  **?** ______________

3. **How many**  **?** ______________

4. **Which has more?**  

5. **Which have the same number?**

Gather together several of your child's favorite toys. Group them according to color. Make a simple bar graph listing the color of toys, and have your child color in the number of bars for each color.

Name ______________________________

Directions

Use the picture graph to answer the questions.
Fill in the bubble to show the correct answer.

				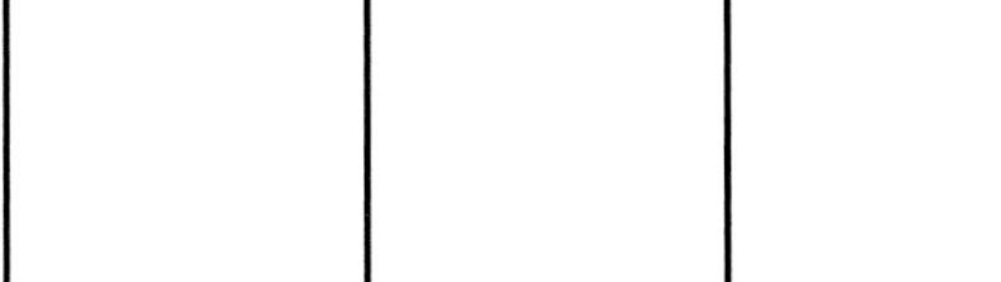		

1. How many  ?

 **2**

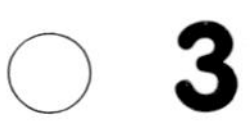 **3**

 **5**

---

2. How many  ?

○ **2**

○ **3**

○ **5**

---

3. Which has more?

○ 

Name ______________________________

Directions

Make a bar graph of the .
Use the graph to answer the questions.

1. How many ? ______________

2. How many ? ______________

3. Which have the same number?